Transducer Handbook

User's Directory of Electrical Transducers

H. B. Boyle

Editor: David Page

BH NEWNES

Newnes
an imprint of **Butterworth-Heinemann Ltd**
Linacre House, Jordan Hill, Oxford OX2 8DP

PART OF REED INTERNATIONAL BOOKS

OXFORD   LONDON   BOSTON

MUNICH   NEW DELHI   SINGAPORE   SYDNEY

TOKYO   TORONTO   WELLINGTON

First published 1992

© Butterworth-Heinemann Ltd 1992

British Library Cataloguing in Publication Data
A catalogue record for this book is available from the British Library

ISBN 0 7506 1194 4

Library of Congress Cataloguing in Publication Data
A catalogue record for this book is available from the Library of Congress

Composition by Page '90

Printed and bound in Great Britain

Cover design courtesy HBM Ltd and Lucas Schaevitz Ltd

Contents

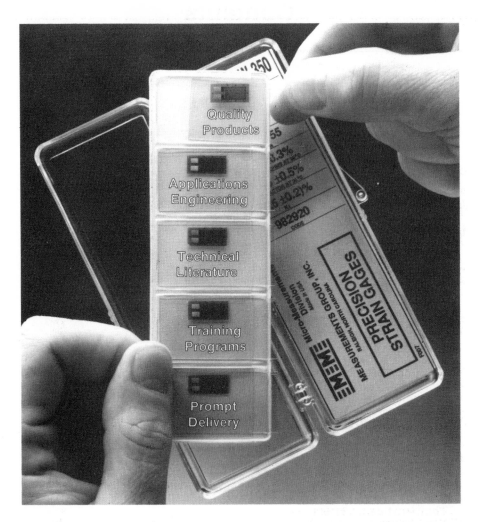

There are five quality Micro-Measurements strain gauges in every package, and more than 25 years of designing and manufacturing know-how in every gauge. Additionally, there's over-the-phone support in applying them, hands-on-training in installing them, and technical literature for understanding them. But, perhaps best of all, there's also our Super Stock commitment for maintaining an extensive selection of the most commonly used Micro-Measurements gauges, so they will be available for immediate shipment whenever you need them.

At Measurements Group *Your* success is our goal!

MEASUREMENTS GROUP U.K. LTD

Stroudley Road, Basingstoke, Hants, RG24 0FW
Tel. (0256) 462131, Fax. (0256) 471441, Telex 858520

Transducer types

Mass, force

Pressure, fluids

Pressure, sound

Relative humidity

Temperature and heat

Time interval

Torque

Velocity, angular

Velocity, linear

Preface

Measurement plays an integral part in the history of the human race. Whenever an ancient language has been deciphered, somewhere within the writings is a description of a measurement and usually a defined standard identifying it. Kings and princes have willingly, or sometimes with a degree of force, signed documents committing them to identify and maintain standards of measurement.

Measurement is inexorably bound up with our everyday life. However in the pursuit of quality measurement there are only two classes of people, namely: those who understand the measurement and ... the vast majority!

There is no doubt that many engineers and scientists are baffled when it comes to choosing the appropriate measuring sensors for measurement, although they seldom admit it.

As a measurement engineer of some thirty-five years experience, the author has repeatedly had to point out the limitations of a selected measurement – why, when a perfectly reliable and accurate measurement has been achieved, the data may be of no value whatsoever!

This book has been written to aid those people willing to seek help and who wish to be aware of the great diversity of electrical transducers.

Acknowledgements

This book would never have been written but for two important factors. First, thanks to all of those engineers and scientists who with patience and kindness taught me my trade; I hope that this book will justify, at least in part, their time and effort. Second, to all of the transducer manufacturers who without exception provided every piece of literature I requested, although in each discussion they were fully aware that the text was likely to be anonymous and that the author was not purchasing any products: thank you. Thanks to my proof readers, Andy and Clive Hudson for reading the text and correcting the errors; and to Joy, my wife, who has increased the "ups" and softened the "downs" of authorship.

Finally, thanks to Bryant Kettle who regularly "saved" information from oblivion when the combined stupidity of the computer and operator resulted in an apparent stalemate!

H.B. Boyle

Amendments and additions to this handbook are always welcome and should be sent for the attention of the author to the following address:

Show and Tell Ltd

204a Thames Side

Laleham

Middx TW18 2JN

UK

A sensible approach to sensing

Making a start

This book is about sensors, or measuring instruments, or transducers – whatever is your preference for names. However to make such a book manageable it has been necessary to impose a limitation. All sensors identified herein have as part of their design, construction and operation an electrical output, and thus the vast majority require an electrical supply.

Even then a further qualification is imposed. Some devices cannot be separated simply into sensors which are then supported with suitable electronics but can only be considered as a complete and often complex system. Distance measuring techniques using radio ranging is such an example and hence large systems are not mentioned here.

Transducers are so called because when they are subject to some physical change they experience some other related change. Only rarely is this physical change something other than a change of position or a deformation.

Because the output is to be electrical, it is not surprising that the vast majority of transducers operate using changes of resistance, inductance or capacitance together with piezoelectric effects, with each competing in markets where its characteristics are presented most favourably. Manufacturers tend to specialise by selecting a single technique to meet the requirement of one type of measurement throughout their whole range. This specialisation provides unexpected examples of ingenuity when techniques are used which at first sight would not appear to be practicable for their chosen application.

The nature of the input supply, when such is necessary, and output signal varies enormously from dc to high frequency, with some requiring further amplification and others not.

The requirement then for inclusion in this directory is that the sensor must produce an electrical output, voltage or current, before any further signal treatment. Some, however, are adaptations of a mechanical device which have been modified to produce an electrical output, sometimes in tandem with the normal mechanical display.

The book is designed as a work of reference and as such is laid out for easy location of each product or measurement type. However, it is commended to be read throughout, particularly for the student/undergraduate, and those who are likely to earn a living making measurements.

In selecting a particular type and range of sensor there are a number of factors which must be considered. It is important to bear in mind that once a choice has been made and the sensor is installed a bad choice not only inhibits the desired quality of measurement but also a corrective re-installation may prove to be an expense far beyond the cost of the sensor itself.

In acquiring data, the aid of many manufacturers has been called upon. Thus, for the more popular devices, the information is the distillation of data from a large number of manufacturers. For some more specialised measurements, the information is based upon a much smaller sample. However it can be asserted that all of the data comes from companies with established records of providing well-developed equipment of a good quality. Please note that the photographs and drawings (which are gratefully acknowledged) are only representative of the products described and will vary in detail between manufacturers from the description given in the text.

1. Measurement Requirement

This requirement always appears to be the easiest part of selection whereas it is here that the greatest number of bad decisions are made.

The problem is that most people do not specify what they want to measure but instead the type of instrument they want to do the measuring. As an example, consider a possible requirement for flow measurement in a pipe. What is the purpose of the measurement? Is the need for instantaneous flow, average flow over a period, or total flow over a considerable period of time? Then, when it is installed, will it interfere with the flow? Only very seldom does a sensor not have any influence upon the measurement. Will the temperature of the fluid be a problem? Is the liquid "clean"? Does it exhibit aeration? Is there any likelihood of solids being present and if so how large are they likely to be?

It is advisable to write down the aims of the measurement keeping the mind blank as to a possible choice of sensor. The measurement will be used for some purpose. Is it clear as to what is that purpose? Is there only one such aim? There is nothing more risky than trying to make a single sensor provide information for two or more unassociated needs.

We have selected "flow" as the example. Is this really the required measurement? Say that the purpose is to protect against an overflow condition. Perhaps pressure or liquid level in the system may be an easier or a better measurement. Be sure that the true requirement is identified.

2. Sources of Inaccuracy

a) Accuracy

In the context of measurement, this is one of the most abused words in the English language.

It is impossible to define accuracy as such, since it must be a summation of all errors, some of which may well not be known with certainty and others not even suspected. Further, the accuracy will be influenced considerably by the actual working range of a transducer compared to its designed range.

It is recommended that the reader seeks information on the meanings of all of the terms which provide information about the efficacy of a sensor. Below are some simple definitions and the associated dangers. Remember *all* of those discussed contribute to the overall accuracy but there are also other factors related to the sensor's surrounding environment and the quality and regularity of calibration. Each has a "sting in the tail" in that the honest definition cannot express the hidden risk in accepting a claimed value without considering what is not defined.

b) Error

This is a much better term to express what is often referred to as accuracy. Its value is the algebraic difference between the measured value and true value of that which is being measured (the measurand). It is usually expressed as a percentage of full range output and must embrace all of the individual components which have been considered in the make-up of the error. Because of the desire to express it as a percentage of full range, since this puts the sensor in the most favourable light, over-specifying the sensor to avoid the risk of over-ranging will affect the error value in proportion.

Note that it is still common to use the term "full scale" rather than "full range" and this arises from past practice where almost exclusively the output was displayed in analogue form on a circular or linear scale. Such display is now becoming less common and hence the preference for "full range" in the text.

c) Hysteresis

Defined as the maximum difference in output between measured values for increasing and decreasing calibration values.

Hysteresis was first identified in magnetic devices where the residual magnetism, say in an increasing magnetic field, opposed the change of magnetism as the values were reduced then reversed in polarity.

Fortunately hysteresis is always larger the greater the measurement excursion. Hence for a sensor cycling over a small part of its range will have a better performance than the full range hysteresis error indicates.

Hysteresis is difficult to determine since a selected value must not overshoot to some higher or lower value until the measurement is made. Additionally the applied input should be a smooth one without drastic changes in the rate of application.

A further problem is that the hysteresis measured includes that of the calibration device and thus errors attributed to the sensor can lie in the calibrator. This error source is often heightened when the two sets of calibration values, increasing and decreasing, are achieved in subtly different ways. For example, an oven used to calibrate a thermometer will be switched ON for the increasing cycle and OFF for the decreasing one. In poor quality ovens, the infra-red emission from the heating elements may heat the sensor in a different manner to the slow cooling procedure. This will give the appearance of hysteresis but is a false representation.

d) Linearity

This is simply a measure of the departure from the output to the sensor for some specified "straight line" law over the working range.

Very few sensors can have a truly linear output. Each change of output results in a changed state so that the next increment of change starts from a new condition. However, the ease with which a sensor output can be manipulated if it is assumed linear ensures that manufacturers seek the designs which give the best linearity possible. There are many occasions where the measurement required is only the difference between the two measurement conditions over a short interval of time. For a linear device, small shifts of the zero value are unimportant, whereas for a nonlinear device significant errors can arise.

Usually expressed as a percentage error of full range output, linearity can be deter-mined in a number of ways. These will give different answers and affect the quoted errors differently for the same set of values. They are all valid, but make sure that the chosen presentation is known.

Often the zero output is ignored in preference to other values when determining the linearity. However, valid though such a technique is, it is important to realise that the output from many transducers when subjected to no physical input will reveal a considerable amount of information about the quality of their performance. Indeed the "zero value" is the only one that requires no calibrated input and thus is the easiest value to obtain. Its long-term study can be revealing, providing information on general stability, over-ranging and electrical noise picked up by the transducer and its associated cabling.

e) Temperature Response

This is sometimes identified as "thermal response".

There are few sensors which do not exhibit some change of output with temperature. Often, internal circuits are incorporated to compensate for this effect. There are two possible sources of error, one of which affects the zero value and the other the range or span of the device. They may require separate treatment depending on the cause. The compensation always assumes that the sensor is at a given uniform temperature and hence if there is a temperature differential across the sensor, the compensation will fail to operate properly and indeed could make matters worse.

For relatively small temperature excursions, the effects may usually be treated as a linear error, but over a wider range some second-order corrections may be necessary.

Some temperature effects can be defined with precision, such as the changes in the elastic modulus of a metal element, which may then allow post-measurement correction. Others are not so readily identified because of the overall complexity of the various components, each making some contribution to the total.

f) Cross sensitivity

This term incorporates all the effects which are likely to influence unfavourably the output of the device other than that due to a temperature shift. It includes the more specific term "transverse sensitivity", or "interaction" when discussing multi-axis measuring devices.

Generally, transverse sensitivity relates to devices which experience and measure a vector quantity such as force or acceleration. It is a measure of the output of a sensor when subject to an input in a transverse direction to that of the chosen one. Again, it is often expressed as a percentage of full range but, as in the example of multicomponent force balances, such comparisons may have no mathematical relevance where, say, force and moment may be under comparison.

Transverse sensitivity must be judged as to its importance in the environment in which it is being used. An accelerometer could well be in a high frequency and high acceleration transverse vibration field whilst trying to measure a low frequency and low acceleration event.

Whereas transverse sensitivity is specific to the type of measurement, that is the comparison of like inputs with only a directional change, cross sensitivity is wider in its definition. For example, a pressure transducer cannot experience a "transverse" pressure as such but may well react to acceleration. Further, poor installation can result in modifying a transducer's performance.

g) Repeatability

This is a term of great significance but which over recent years has tended to become degraded by the manner of its usage.

Repeatability is no more than the capability of a sensor to repeat its performance in two or more tests. Again it is usually expressed as a percentage of full range output.

The danger lies in the fact that sensors may exhibit satisfactory repeatable performance when subject to a single selected test environment but once disturbed and reintroduced to the same environment, may produce a significant change in output. This leads to a new term which is gaining increasing credence and acceptance, namely, "reproducibility".

h) Reproducibility

In this type of test there is a deliberate policy to remove and reinstall the device between at least two test runs. As a simple rule, it is advisable to proceed as if the device and its associated electronics were being prepared for packing or storage. Thus the electrical input is disconnected, any test attachments are removed, plugs and sockets disconnected and, most importantly, the transducer is removed from the calibrator. The effects of such removal can be quite startling!

i) Resolution

This is a measure of the step change of output signal of a sensor when experiencing a continuous physical input change. In some ways it is an archaic term since nearly all sensors today may be described as having "infinite resolution".

However, this term, in its turn, may be somewhat misleading, as there will be a "smallest" increment which it is possible to measure whatever the resolution. This may be influenced by the smallest increment available for display on the amplifier system or the limitation imposed by the value of the electrical "noise" generated by the sensor and electronics combined. A further limitation to the resolution capability is becoming more common as data are being handled by digital means. For example, if data are digitised in an 8-bit processor then the resolution can be no better than 1 in 256 parts, irrespective of how many numbers are displayed.

j) Frequency Response

This is possibly the most neglected of those parameters which have an influence on the quality of measurement. In part, this arises from the method of calibration. The aim of calibration is to subject the sensor to a steady input condition, allowing time for such a condition to be reached and then to note the output response. Seldom is the sensor subject to an input variation as part of its calibration. (Even when the input is cyclic, the selected frequency will be maintained constant for each test condition.) Hence, the master calibration may not include the frequency response and this, if established at all, is likely to be a separate test.

Although it is extremely rare that during calibration the input is other than constant, in use it is rarely so. If the input oscillation is of sufficiently high frequency, the sensor can change its state and measure something entirely different. For example, an accelerometer driven above its natural frequency begins to exhibit characteristics more akin to a displacement transducer. Again, although temperature is one of the easier measurements, most thermometers respond poorly to rapid temperature changes and even the term "response time" can be misleading since there are different interpretations of this value.

Manufacturers do quote natural frequencies and of course for some measurements this is important and for devices such as accelerometers, such information is essential. However, it is necessary to be sure of the test conditions which were used to identify the frequency. A pressure transducer used for gas pressure measurement will have a higher operating frequency than when measuring liquid pressure since the liquid will influence the mass of the vibrating system.

9

k) Transducer Interference

This error source is never mentioned in the catalogues or indeed standard reference books. However, there is an ever-present risk that a sensor will experience a change in output which does not arise from the chosen input due to some unidentified external influence.

Friction is perhaps the most common example of this problem, but there are others. The incorrect installation of a sensor, overtightening of a component, cross energising of the sensor electrical supply and capacitive or inductive coupling are some of the common sources of the problem.

The nature of the interference is that, remarkably, the error can be quite systematic and is "lost" in the data and the information is accepted as "valid". The only protection is to be vigilant and make as many individual checks as time and money will allow.

3. Calibration

To some, calibration is a precise, carefully-executed activity. To others it is little more than a necessary delay in proceeding with the measurement programme. Rarely is calibration cheap. Sometimes it may cost more to perform than the value of the sensor. It will seldom, possibly never, be undertaken in a manner which matches the environment of use.

There may not be a defined national or international code or standard method of calibration. So how does one overcome these technical hurdles? It is possible to make some general recommendations as follows:

i) When a calibration is defined by a written Standard then read the specification with care to find out what is and what is *not* included in the defined calibration. For example, BS 1610: Part 2, Materials Testing Machines and Force Verification Equipment, does not require the load-measuring device to be subject to a temperature range; indeed it specifies that the temperature must not vary by more than 1°C during a test cycle. Hence a calibration may well be achieved on a load cell which has more to offer as a thermometer! This is not to be critical of this particular Standard since here the document is more concerned with the temperature errors which are to be corrected in a proper manner during analysis of the results and subsequent use, which can be achieved only if the calibration temperature is maintained as steady as possible.

ii) Compare the specification environment with that to which the sensor will be subjected when in use. From this deduce the likely additional errors which could be present.

iii) Always ensure that at least three complete tests are undertaken to define the calibration. Use the "reproducibility rule" above at least once.

iv) If the calibration is being undertaken in-house, ensure that all of the equipment is traceable to national or international standards. Just because a device displays its output in engineering units it does not follow that these are correct. Always make an honest estimate of the likely errors present in the calibration equipment; a sensor cannot be better than the sum of its own errors and those of the calibrator.

There are many occasions when the assurance of a detailed calibration may be suspended with perfect safety. Here, reliance is placed upon the supplier. Reputable suppliers, however, will provide a calibration as part of the overall package. There are also a wide range of sensors which, because of their design and quality control, will have identical performance, and thus a calibration of one transducer will serve a whole family of devices. Cup anemometers are an example of the latter. When such a bulk approach is used, then the claimed performance will of necessity and circumspection be of a lower order, but will often serve the client's need. However, this approach does require that the sensor be maintained and serviced as defined by the manufacturer and with the use of specified replacement components.

4. Signal Conditioning

This book concentrates on sensors. This does not mean that signal conditioning, amplification, filtering or other signal manipulation can be separated from the complete system. Indeed the two parts, sensor and signal conditioning, must be treated as a whole.

The signal conditioning may well suffer from at least some of the errors discussed above. There is a tendency to ignore these, but do so at your own peril! Zero stability, supply voltage variation, gain, reproducibility, linearity and resolution are the more obvious errors for consideration. As an example, many amplifiers have a greater temperature sensitivity than the sensor attached to them.

Read the specifications with great care and ensure the quality of the signal conditioning is a proper match to the sensor. It may not be necessary to buy both the transducer and signal conditioning from one manufacturer, but, by so doing, it does offer the protection that when there is a known problem, the complaint may be made to one company rather than setting up the potential for a three-way argument.

5. Data Analysis and Display

The analysis of the measured quantity can be as simple as making a mental note of the values, such as reading a fuel gauge on a motor vehicle, or as complex as data analysis supported by a computer. The detail of this analysis will often influence the sensor selection. The fuel gauge hardly requires a high order of accuracy if it is used only as a reminder to "fill up soon". However, for the development engineer there can be a definite requirement to measure consumption to a defined precision.

Thus it is important to think carefully about the use for which the data has been acquired. Indeed, there is one approach which suggests that data analysis is the place from which to start when a measuring system is being proposed. This can become quite detailed and include "dummy" analysis so that maximum values can be predicted. It is important to remember that all measurement is the result of the need to confirm a prediction or to display a warning.

Returning to the fuel gauge, in the development situation the maximum likely value of fuel flow has to be considered, whereas for the average motorist, concern is more about the possibility of being stranded with an empty fuel tank. Choose the gauge with care so that the solution fits the need!

6. Intrinsic Safety

There are many requirements where the transducer may offer a threat to its local environment and thus some suitable protection must be offered. Electrical devices, with their potential to provide spark ignition, are especially in need of scrutiny to ensure adequate protection. There are many specifications which identify the proper proce-dures but such protective measures may influence the choice of transducer and degrade its performance. It it worth seeking professional advice before purchasing equipment.

The Directory

The detailed information contained within the bulk of this book is in the style of a directory and is designed to help the reader select the correct sensor for the purpose. To describe a whole family of sensors in the space of a few hundred words requires that descriptions can be no more than brief, hopefully succinct, statements.

When outlining any type of sensor there will be some devices which have not been included because of their very specialist use. What is attempted here is to identify that wide range of sensors which are the common choice, and not to dwell on the exception.

Below is the interpretation of the various headings used throughout the book. Where numerical data are provided the information may not be related to any particular transducer. Thus it may not be possible to obtain a given quoted error to the identified temperature range nor a range spread from a single source manufacturer.

The order in which each characteristic for a particular type is presented is based upon the logic and convenience of the layout and is in no priority order. Readers should however decide upon their own priorities. For example, "over-ranging" is often more important than "error" since there may be little point in having an extremely accurate transducer which never makes a measurement because it fails in an over-ranged condition.

The link to any section is the units which define the measurement. Thus "length" leads to "displacement", "extension", "position" and "strain". These groupings do not necessarily relate to each other in true mathematical terms. Indeed in the examples quoted above, three are expressed in terms of length and one is non-dimensional. However to the measurement engineer, experience shows that possibly all of these methods could be in contention when considering a particular length measurement problem; whether for displacement, extension, strain or stress determination.

An Introduction

This precedes all sections of the Directory and is a discourse on the type of measurement and a general discussion of the various options. To some extent these comments are a matter of personal opinion. They are offered in good faith and are based upon considerable expertise of not only the author but also of other measurement engineers.

A brief indication is given on the type of calibration which can be achieved by using possible in-house equipment.

At the end of each introductory discussion, the NAMAS accreditation laboratory number is given for those seeking an independent external calibration service. This also provides an indication of the availability or otherwise of calibration services for a particular type of measurement or transducers. The information has been obtained from the NAMAS Concise Directory (M3) August 1990.

For those who wish to use NAMAS-accredited services, the Directory can be obtained by writing to:

NAMAS Executive,

National Physical Laboratory,

Teddington,

Middlesex TW11 0LW, UK

The Directory Format

Information relating to over 100 sensor types is given in a consistent style for ease of reference, one type per page. At the risk of being repetitive, the operational descriptions are given for each variation within a type to ensure that the reader does not have to refer to other pages. (The wording may not be precisely the same since the emphasis may vary slightly from type to type.)

There are, of course, a considerable number of attributes which may not be mentioned including water immersion proofing, nuclear radiation resistance or explosion-proof capability. These qualities need to be reviewed with the manufacturers since a discussion of specialised protection requirements would take another book.

In providing information about the various sensors, the quoted values may be described as "typical best performance", that is for any set of values there may well be a sensor which improves upon the quoted values and an even wider choice of others which are worse.

The units used are those normally accepted in the SI system, with acceleration being the exception. Note that, since some values are conversions from other unit systems (such as Celsius from Fahrenheit) these values may give the impression of greater precision than intended.

a) Transducer Type

This identifies a particular type of transducer and includes a code which is unique to that type. This code, together with supplier codes discussed below, allows cross referencing in the list of suppliers and addresses given in the appendix.

b) Operating Principle

This gives a brief, sometimes simplified, description of the way in which the device operates. There is a wide range of sensors with commonality in their mode of operation such as in the use of bonded resistance strain gauges for measurements of force, pressure, displacement and acceleration.

c) Illustrations

These are included to show typical shapes and types of device.

d) Application

This section is a summary of typical uses for a particular type of transducer. It is not exhaustive nor does it infer that the transducer has no other application.

e) Range Limits

The word "range" is preferred to "span" and this entry identifies the readily available ranges from a typical minimum, to maximum of full range capability. Note that not all transducers operate down to a zero (static) input condition.

Where important, other information such as the frequency range is also supplied here.

f) Error Level

There is some variation between devices as to which of the errors discussed previously are included. Hence, where possible, the error sources are identified either individually or in combination to achieve the quoted values.

It is still common for manufacturers to use "accuracy" in their description of performance. Whilst not giving approval to the use of the word, one can have sympathy with the problem. There are transducers where the identification of specific components of error are extremely difficult to establish. However, the manufacturers have undertaken considerable calibration to ensure the performance. Hence "accuracy" is included on occasion with no identification or distinction of the error sources.

Some types of transducer come supplied complete with an integral power supply and signal conditioning system and are energised with either a mains supply or unregulated low voltage dc supply. For these devices the error values are given for the whole system.

g) Over-range Protection

Nearly all transducers are at risk from some form of over-ranging. The values are sometimes quoted in two parts, namely that which can be accepted without damage and that which can be accepted without total structural failure.

Sometimes the ability of a sensor to withstand over-ranging can be its most important feature. Some devices are essentially immune from this and when so identified the text considers other factors which may result in failure.

Because a transducer can accept an over-range condition, it does not mean that such continuous exposure is acceptable since long-term over-ranging may still result in premature failure.

h) Supply Voltage

This specifies the electrical input to the transducer, both voltage and frequency. For some transducers, the power supply and signal conditioning form an integral part of the device and when they do so, the supply voltage for the system is quoted. Where this is from a conventional power socket it is identified as "mains".

i) Sensitivity

This is a measure of the output signal and may be expressed in a number of ways. Sometimes it is expressed as a ratio of the output voltage to the input supply for full range excursion. For others it will be the output voltage per unit of the measurand and sometimes no more than the full range output. Whatever is used is made clear in the text. For sensors with outputs in the form of a voltage level, it is assumed to be input to a high impedance source since changes in impedance may have an effect upon this value.

j) Cross sensitivity

This is an important parameter that is not always readily available from the manufacturer. It is preferred to use "cross" rather than "transverse" for this reason. Transverse sensitivity is usually understood as the extent to which the measure of the transducer can respond to a similar parameter in a direction perpendicular to the desired one. For example, an accelerometer will be designed to respond along a selected directional axis. Transverse sensitivity is its unwanted response to acceleration along another axis.

Some transducers do not experience such effects – pressure is such an example. Where transducers may be sensitive to some other undesirable effect, this is described. In the case of the pressure example, this could be the effect of vibratory acceleration upon the diaphragm. Other types of transducer are extremely sensitive to their environment or the method of mounting and thus these different effects have been collected together under the wider-ranging term of cross-sensitivity.

k) Temperature Range

There are three possible temperature ranges which may be quoted. They are:

Compensated range, where some extra temperature sensing components have been included to reduce or cancel the temperature error;

Operating range, which identifies the full temperature range and will encompass any identified compensated range;

Storage range, which is the non-operating range and must encompass the other two ranges above.

The temperature quoted may distinguish between these three options. Where they are not so identified it should be assumed that the quoted values are coincident to all three.

l) Temperature Sensitivity

Sometimes called thermal sensitivity. Depending on the construction and type of sensing, transducers may be designed so that there is an automatic compensation opposing the effects of temperature change, whereas for others a dedicated temperature compensation circuit may be incorporated. Even when transducers are fitted with temperature compensation there will be some residual effect. This comes in two forms:

Zero shift. This is the amount by which a transducer output will change when subject to a change of temperature alone. Although this effect is usually independent of the actual measurement state, it is easiest to establish the value when the transducer is under "zero" conditions. (It is sometimes believed wrongly that this error is only present at the zero condition whereas it is present whatever the state of the transducer.)

Range or span shift. This is a measure of the range change caused by the effects of temperature. This will usually include the change in the elastic modulus of the "spring element" of the transducer.

When it is specified that a sensor is temperature-compensated, the temperature sensitivity applies to this temperature range alone. Where no such compensation is declared, the temperature sensitivity applies to the full operating values.

m) Signal Conditioning

It would be improper to describe a sensor without drawing attention to the type of signal conditioning that accompanies it. Whilst transducer designs change relatively slowly, the effects of modern electronics and data processing have led to a dramatic revolution. Thus items identified here are likely to be minimum requirements.

n) Benefits and Limitations

This will help potential users to select the best sensor for the required task. The comments must be partly subjective because there will be exceptions in the general suitability of a particular type of transducer to perform an allotted task. Factors which may well transcend the general performance, such as size, simplicity of installation, immunity from electrical interference or low cross sensitivity, are just some which may have the most significant influence on choice.

o) Comments

Under this heading is additional information. These comments are very much personal opinions and thus may conflict with other opinions. However, they are offered in good faith and are based on the author's own experience.

p) Suppliers

The codes here identify the suppliers listed in the List of Suppliers.

Acceleration

A cceleration measurement is of vital importance to the engineer since the life of a structure or artifact often depends upon the severity of the acceleration it experiences during vibration.

Accelerometers are devices which rely upon a mass/spring system, preferably of a high natural frequency. When the system experiences a vibration, the mass, more precisely the seismic mass, exerts a load upon the spring and thus by measuring this deformation, or some representation of it, a measure of the acceleration is made.

Although it is possible to purchase **angular accelerometers** as described here, on occasion it may be sufficient to use matched pairs of **linear accelerometers** separated by a suitable distance and operating in parallel. By algebraic manipulation a measure of both the linear and angular acceleration is obtained; the accuracy will however be lower since two devices are used.

The performance of accelerometers is influenced by their own natural frequency, the damping ratio and the manner by which the damping is achieved. This in turn affects the amplitude and frequency response of the device within the required operating band.

Thus for a complex vibration, arising from some fundamental frequency and a number of related harmonics, the phase relationship measured may require appropriate correction. Even where the absolute relationship may not be needed, if the requirement is for relative measurement between two or more locations it may be best to ensure that only one product of a selected manufacturer be used which will give a common error.

To protect the transducer from excessive motion of the spring/mass system which arises as an accelerometer approaches its own natural frequency, damping may be incorporated into the system with the upper frequency of operation specified well below the natural frequency. However not all accelerometers are provided with damping which imposes an even lower operating frequency. This, however, provides a better frequency/phase relationship.

Acceleration, linear

The choice of the type of accelerometer for a particular activity must be guided by the nature of the requirement.

Continuous supervision of vibration may be better with one device whilst intermittent usage as in "trouble shooting" applications may suggest a different device. Again, many problems are related to frequency alone. For example, if a structure is failing due to self-generated vibration then the search for the source of the vibration input will certainly require a frequency-related component whose amplitude characteristics are of less or little import.

Commonly an accelerometer's electrical output may be converted into either velocity or displacement, mathematical manipulation being achieved by electronic means. This technique is used for displacement where no fixed reference can be defined. As an example it is used to determine wave height from a freely moored buoy. Here an accelerometer is maintained vertical and the "displacement" signal is transmitted by radio to a shore-based receiver.

Unlike other transducers identified in the text, accelerometers are always quoted in terms of their "g" value rather than some scientific unit. This tradition is maintained here. (1 g = 9.81 m/s^2 approximately).

In selecting a transducer for a particular application, acceleration and frequency range are not the only criteria for consideration. Indeed, acceleration levels in a vast number of engineering applications are likely to be lower than 10 g with the natural frequency of the structure lower than 30 − 40 Hz with frequencies below 10 Hz being common. In such cases the choice is more likely to be influenced by other factors such as its physical size.

For lightweight structures there is a risk that the increased mass of the attached accelerometer will lower the natural frequency of the original structure to some unacceptable level. Robustness too, including general protection from the environment, is a major factor when selecting a transducer for applications in hostile environments. Indeed this type of application is just one reason why piezoelectric transducers are often favoured.

Calibration of accelerometers may be achieved in-house by one of three methods. For values below 1 g, the accelerometer may be placed upon a horizontal surface and then tilted through a number of known angles, each angle representing some percentage of the vertical component of the Earth's gravitational field.

Alternatively the accelerometer may be placed upon a horizontal disc which is then rotated. A knowledge of the radius and speed of rotation determines the acceleration component. These two techniques are suitable for devices which respond to steady acceleration and for the latter technique high g values are possible.

For devices which respond only to dynamic conditions or for very high acceleration levels some form of vibrating table or "shaker" is required. This now leads into the problem of establishing the amplitude of vibration. Unless proper equipment is to hand this last technique should be avoided for in-house design.

NAMAS Accredited Laboratories: 0060, 0151 and 0273.

Contents: Acceleration, linear

AC1. Bonded resistance strain gauge

±1 g to ±500 g.

AC2. Capacitance

±10 g to ±30 g.

AC3. Electrolytic

±0.1 g to ±1.0 g.

AC4. Inductive

±1 g to ±120 g.

AC5. Piezoelectric

±5 g to ±100 000 g.

AC6. Piezoresistive, bonded semiconductor strain gauge

±2 g to ±2000 g.

AC7. Piezoresistive, integrated semiconductor

±2000 g to ±200 000 g.

AC8. Servo or force balance type

±0.25 g to ±50 g.

Acceleration, linear

AC1. Bonded resistance strain gauge

OPERATING PRINCIPLE These devices comprise a mass and metallic spring element. The spring element has bonded to it strain gauges, usually to form a conventional Wheatstone bridge, and thus when subject to vibration the spring is deformed and the resulting strain provides an output directly related to the applied acceleration. Some devices come complete with the signal conditioning incorporated.

APPLICATIONS A good general-purpose accelerometer capable of many vibration measurement tasks. For hazardous applications it is generally classified as a "simple" device and hence can be protected by zener barriers.

Courtesy Graham & White Instruments

RANGE LIMITS ±1 g to ±500 g[1]. 0 – 200 Hz to 0 – 2000 Hz.

ERROR Non linearity and hysteresis: ±1% of full range.

OVER-RANGE PROTECTION Mechanical stops may be fitted to protect against overload. A likely minimum is ×5 of the operating range but higher values are common.

SUPPLY VOLTAGE 1 V to 6 V and unlikely to exceed 10 V.[2]

SENSITIVITY Dependent upon range with 2.5 mV/V for full range a likely maximum.

CROSS SENSITIVITY Transverse sensitivity values of ±5% of full range or better are obtainable.

TEMPERATURE RANGE Compensated: + 15°C to 93°C. Operating: – 40°C to 93°C.

TEMPERATURE SENSITIVITY Zero: ±0.01% of full range per °C. Range: ±0.02% of full range per °C

SIGNAL CONDITIONING Most manufacturers supply the appropriate signal conditioning. Alternatively strain gauge or differential dc amplifier with an auxiliary power supply are suitable.

BENEFITS AND LIMITATIONS [1]These sensors come in a good variation of ranges and physical sizes with the higher frequency responses related to the higher ranges.

COMMENTS Attachment is usually a simple "bolt down" system using two or more fastenings. Being resistive devices they can be energised from ac power supplies but as a general rule the operating frequency is then limited to at least 1/5 and preferably 1/10 of the ac energising frequency. [2]Some devices incorporate the signal conditioning within the transducer housing. Low voltage such as 28 V dc is typical.

SUPPLIERS GRA; PRE; RDP.

Acceleration, linear

AC2. Capacitance

OPERATING PRINCIPLE The motion of the mass is controlled and restrained by a suitable spring arrangement. The mass incorporates two separate capacitor plates and vibrates between two other similar plates which are rigidly attached to the transducer housing, thus forming two separate capacitors. These are electrically connected to form a half bridge capacitor circuit and the resulting differential change in capacitance is a measure of the applied acceleration.

APPLICATIONS Include aircraft, missile and automobile test applications. The extremely high shock capability combined with the capability to operate in high radiation environments suggest other uses.[1]

Courtesy Endevco (UK)

RANGE LIMITS ±10 g to ±30 g. 0 – 600 Hz to 0 – 1000 Hz.

ERROR Non linearity and hysteresis: ±0.2% to ±1.0% of full range subject to selection.

OVER-RANGE PROTECTION Sinusoidal: ×100 g. Shock: 10 000 g.

SUPPLY VOLTAGE Requiring a low voltage energisation such as 13 to 18 V dc.

SENSITIVITY Full range: ±2 V output from incorporated signal conditioning.

CROSS SENSITIVITY Transverse sensitivity: ±1% of full range.

TEMPERATURE RANGE Operating: –55°C to + 121°C. Storage: –73°C to + 150°C

TEMPERATURE SENSITIVITY Zero: ±0.02% of full range per °C. Range: ± 0.02% of full range per °C.

SIGNAL CONDITIONING These transducers are supplied complete with an inbuilt stable power supply and amplifier.

BENEFITS AND LIMITATIONS Capacitance devices have a good reputation for operating in strong magnetic and high nuclear radiation fields.

COMMENTS [1]These are specialist devices generally targeted at tasks such as motor vehicle crash testing. The use of capacitance to measure acceleration is not as common as one would expect and yet performance is comparable with other devices and these transducers can measure down to zero acceleration levels.

SUPPLIERS END1; END2.

Acceleration, linear

AC3. Electrolytic

OPERATING PRINCIPLE An electrolytic liquid is restrained in a curved tube much like a bubble spirit level. Two metal probes, inserted from each end, pass through the liquid and into the bubble. A third fully submerged probe is also inserted. When the bubble is central, the electrical paths between the first two probes and the third one measure identical resistance values. When the bubble moves there is a resultant resistance differential between the two measurements.

APPLICATIONS These devices are simple and robust. Hence for low horizontal accelerations these devices are excellent.
Uses include military, civil and structural engineering applications.

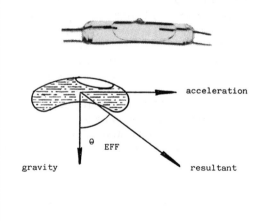

Courtesy IFO International (UK) Ltd

RANGE LIMITS ± 0.1 g to ± 1.0 g[1].

ERROR Non linearity: $\pm 1\%$ of linear range. Asymmetry: $\pm 2\%$ to $\pm 5\%$ of total range.

OVER-RANGE PROTECTION Extremely robust. Over ranging will provide nonsensical outputs which will then recover. Shock loadings up to 300 g are possible.

SUPPLY VOLTAGE 6 V to 12.5 V ac, 50 Hz to 20 kHz[2]

SENSITIVITY Discriminations of between 0.00001 g and 0.0001 g are quoted.

CROSS SENSITIVITY Transverse axis tilt: 1 minute of arc per degree of transverse tilt or better are typical for the smaller ranges.

TEMPERATURE RANGE $-10°C$ to $60°C$.

TEMPERATURE SENSITIVITY Zero: Between ± 0.3 and ± 3 seconds of arc/°C. Range: Between -0.1 to $+2\%$ of full range/°C.

SIGNAL CONDITIONING The three probes and two resistors form a Wheatstone Bridge energised by a stable ac supply. Amplification may not be needed.

BENEFITS AND LIMITATIONS An extremely robust and simple device which is capable of providing a measurement of very low horizontal accelerations.

COMMENTS [1]The acceleration values quoted above are for an unambiguous but non-linear range. The linear range is in the order of 1/10 of these values.
[2]The frequencies quoted are the natural frequencies and can be modified considerably by the choice of damping. These devices are equivalent to mechanical pendulum accelerometers. Thus they have an effective pendulum length which determines their natural frequency.

SUPPLIERS TIL.

AC4. Inductive

OPERATING PRINCIPLE A mass is suspended and restrained by spring elements. The displacement of this mass/spring system is sensed by two iron cored inductance coils located on either side of the mass. These two coils form adjacent arms of an ac bridge circuit and displacement of the mass is sensed by the inductances such that the bridge output is proportional to the acceleration.

APPLICATIONS Inductive devices have a very good long term stability and a preferential low "g" range. Aircraft, helicopter, ship and motor vehicle research are typical applications.

Courtesy Lucas Schaevitz Ltd

RANGE LIMITS ±1 g to ±120 g. 25 Hz to 325 Hz.[1]

ERROR Non-linearity and hysteresis: ±0.25% to ±0.5% of full range.

OVER-RANGE PROTECTION ×1.2 to ×5 of full range. Shock: 100 g.

SUPPLY VOLTAGE Typically 10 V at a frequency of 5 kHz to the transducer.[2]

SENSITIVITY 25 mV/V ac to 50 mV/V ac. or ±3 V dc with integral electronics.[3]

CROSS SENSITIVITY Transverse sensitivity: ±0.0034 g/g input.

TEMPERATURE RANGE Operating: –40°C to 80°C. Storage: –50°C to 100°C.

TEMPERATURE SENSITIVITY Zero: ±0.02% of full range/°C. Range ±0.04% of full range/°C.

SIGNAL CONDITIONING An oscillator, transformer and phase sensitive demodulation to provide a dc output.

BENEFITS AND LIMITATIONS This is an ac system so care must be taken to avoid electrical pick-up to and from other sources.

COMMENTS [1]Dependent upon range; the higher the range the greater the frequency of operation.
[2]Note that other energising frequencies may be used but this can have influence on the maximum measurement frequency of the transducer.
[3]Transducers can be supplied complete with all of the necessary signal conditioning within the transducer casing. These are energised by a wide option of input voltages and seldom require further amplification.

SUPPLIERS LUC1; LUC2; SMI.

AC5. Piezoelectric

OPERATING PRINCIPLE Using the properties of piezoelectric crystals to generate a charge on the crystal faces when subjected to deformation, a ceramic crystal element is rigidly mounted into a housing. Thus when it is deformed an electrical charge appears upon the face of the disc. This electrical charge is converted to a voltage by a charge amplifier. The disc provides both the "spring" and a significant part of the "mass" of the system.

APPLICATIONS Used in a wide range of dynamic applications, particularly at high frequencies and where over-ranging is likely. Includes vibration of rotating machinery, structures and shock investigations.

Courtesy Endevco (UK)

RANGE LIMITS ±5 g to ±100 000 g. 1 Hz to 30 000 Hz.

ERROR Nonlinearity: ±0.5% of full range.[1] Hysteresis: Negligible

OVER-RANGE PROTECTION ×10 to ×2 of full range as the range values increase.

SUPPLY VOLTAGE Nil. Self generating.

SENSITIVITY ±45 pC/g to ±1000 pC/g[2]. 1000 mV/g to 1 mV/g[3].

CROSS SENSITIVITY Dependent upon design factors such as ruggedness but the transverse sensitivity is unlikely to exceed ±5% of full range.

TEMPERATURE RANGE Operating: –50°C to + 120°C

TEMPERATURE SENSITIVITY Temperature coefficient: –0.06% of full range output/°C.

SIGNAL CONDITIONING Requires charge amplifiers peculiar to piezoelectric devices alone and further requires special cabling to connect the sensor to the amplifier.

BENEFITS AND LIMITATIONS A wide choice of devices especially in the higher ranges. However they are suitable for dynamic use only.

COMMENTS They are available in a range of sizes, some of which are extremely small. Mounting is generally simple such as an integral stud or through hole mounting. A good knowledge of the way in which the transducer and supporting charge amplifiers work is beneficial to ensure properly installed systems.
[1]Although this value is typical for full range, often the same numerical value will be achieved for partial range calibration.
[2]At crystal face.
[3]After typical signal conditioning.

SUPPLIERS END1; END2; HBM1; HBM2; KIS1; KIS2; KIS3; KIS4; PRE; RDP; TEC.

AC6. Piezoresistive, bonded semiconductor strain gauge

OPERATING PRINCIPLE Semiconductor strain gauges are bonded to a metallic spring element such as a cantilever beam which has a mass attached to its free end.

The resulting spring/mass system forms the heart of the accelerometer and thus when subject to vibration provides an output directly related to the applied acceleration.

APPLICATIONS Easily mounted either by an integral screw mounting or even bonded into position. Typical uses include aerospace research and other lightweight structure investigations.

Courtesy Entran Ltd

RANGE LIMITS ±2 g to ±2000 g. 0 – 60 Hz to 0 – 4 kHz.

ERROR Non-linearity and hysteresis: ±1% of full range or better.

OVER-RANGE PROTECTION Mechanical over-range stops may be fitted and between ×2 and x5 is common with much higher capability for shock loading.

SUPPLY VOLTAGE Between 7.5 and 15 V dc.[1]

SENSITIVITY 20 mV/g to 0.025 mV/g or 75 mV to 250 mV full range.[2]

CROSS SENSITIVITY Transverse sensitivity: ±3% to ±5% of full range.

TEMPERATURE RANGE Operating: –40°C to +121°C.[3] Storage: –54°C to +121°C.

TEMPERATURE SENSITIVITY Zero: ±0.02% of full range/°C. Range: ±0.05% of full range/°C.

SIGNAL CONDITIONING Some devices are supplied with an integral signal conditioning. Where they are not supplied, conventional strain gauge amplifier systems can be used.

BENEFITS AND LIMITATIONS They provide a large voltage signal combined with extremely small size. (Example, weight 0.5 grams, volume 100 mm³).

COMMENTS [1]Semiconductor devices should be operated at the manufacturer's recommended voltage only, otherwise the temperature specification may be altered.
[2]Some devices, with integral amplification, provide ±5 V output for full range.
[3] "Compensated" ranges usually span approximately a selected 50°C band such as 120°C to 170°C. This is a function of the nonlinear properties of semiconductor gauges. Biaxial and triaxial devices are available.

SUPPLIERS END1; END2; ENT; KUL1; KUL2; RDP.

Acceleration, linear

AC7. Piezoresistive, integrated semiconductor

OPERATING PRINCIPLE These devices are similar to the bonded semiconductor type but for very high g capability it becomes necessary to use the gauge element as the spring of the spring/mass system. A silicon chip containing integrated semiconductor gauge elements combine these two functions, spring and mass, in one integrated unit; providing a high acceleration and high frequency capability.

Courtesy Endevco (UK)

APPLICATIONS Lightweight and of small size these devices are mounted by bolt down or integral stud fixing. Designed specifically for high acceleration uses such as vehicle testing.

RANGE LIMITS ±200 g to ±200 000 g. 0 – 10 kHz to 0 – 150 kHz

ERROR Nonlinearity and hysteresis: ±1% to ±2% of reading.

OVER-RANGE PROTECTION ×5 to ×1 dependent upon range.

SUPPLY VOLTAGE 10 V dc typical[1]

SENSITIVITY 15 μV/g to 1.5 μV/g. Up to 0.3 V for full range output.

CROSS SENSITIVITY Transverse sensitivity: ±5% all axes.

TEMPERATURE RANGE Operating: −18°C to + 66°C Storage: −54°C to + 121°C.

TEMPERATURE SENSITIVITY Zero: ±0.12% of full range/°C.[2] Range: ±0.12% of full range/°C.[2]

SIGNAL CONDITIONING Some devices are supplied with an integral signal conditioning. Where not supplied, conventional strain gauge amplifier systems can be used.

BENEFITS AND LIMITATIONS Their extremely high acceleration capability and good signal output is combined with small size.

COMMENTS Very much a specialised device for high acceleration level measurement but with the capability of operating down to steady state levels.
[1] It is important to operate these devices at the manufacturer's stated values or the temperature compensation may be invalidated.
[2] "Compensated" ranges span approximately a 50°C band such as 20°C to 70°C.

SUPPLIERS END1; END2; ENT; KUL1; KUL2.

AC8. Servo or force balance type

OPERATING PRINCIPLE A mass, forming the "bob" of a pendulum, senses the applied acceleration. The resulting displacement of the mass is detected by a position sensor which produces an electrical signal. This signal, in turn, is converted into a negative feed back one which is used to generate a corrective moment to the pendulum arm which then restores the mass to its original position. The signal is also used as a measure of the acceleration.

APPLICATIONS Used where accuracy, especially at low "g" values, is of the utmost importance. Further they are capable of having a full range below 1 g. Mounting is by bolt down fastening.

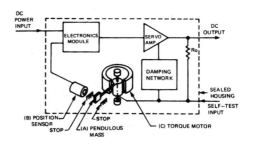

Schematic of Servo Accelerometer

Courtesy Lucas Schaevitz Ltd

RANGE LIMITS ±0.25 g to ±50 g.[1] 30 Hz to 200 Hz.

ERROR Nonlinearity: ±0.05% of full range for 5 g full range or below. Hysteresis: ±0.02% of full range, all ranges.

OVER-RANGE PROTECTION Acceleration typically up to 100 g all three axes.[1]

SUPPLY VOLTAGE Most common low voltage dc power inputs.

SENSITIVITY ±5 V dc, full range.

CROSS SENSITIVITY Transverse sensitivity: ±0.002 g/g.

TEMPERATURE RANGE Operating: –55°C to +95°C. Storage: –65°C to +105°C.

TEMPERATURE SENSITIVITY Zero: 0.002% of full range/°C. Range 0.02% of full range/°C.

SIGNAL CONDITIONING The signal conditioning is incorporated within the transducer casing. With the high dc output further amplification is seldom required.

BENEFITS AND LIMITATIONS Closed loop systems allow the damping to be controlled electrically providing much greater control over the design parameters.

COMMENTS [1]To accommodate vertical acceleration, a built-in bias of 1 g can be incorporated within the device.
[2] Shock loading protection depends upon particular type and design application but values of at least 100 g applied for 11 ms are quoted.
[3] These devices are ac in operation but being a feedback system the necessary electronics are incorporated within the transducer housing.

SUPPLIERS LUC1; LUC2; SMI

Acceleration, angular

The measurement of angular acceleration is a less common activity than its linear counterpart and this is reflected in the limited choice of transducers. There are however many applications used in vibration measurement where it is essential to determine the angular acceleration without the need to identify the rotational axis of that motion. An example is for use on ships or aircraft to determine roll acceleration. Although such motions could be determined by tandem linear accelerometers these are inherently less accurate and any flexibility within the structure may result in erroneous values being obtained.

"Steady state" calibration is achieved by using a rotating table.

NAMAS accredited laboratories: 0060 and 0151.

Contents: Acceleration, angular

AA1. Piezoresistive

±5000 radians/s^2 to $\pm50\,000$ radians/s^2.

AA2. Servo or force balance type

±200 radians/s^2 to ±1000 radians/s^2.

Acceleration, angular

AA1. Piezoresistive

OPERATING PRINCIPLE A spring/mass metallic element is designed to deform when subject to angular acceleration in a chosen plane. This deformation is detected by semiconductor gauges bonded to the metal surface.

The spring element will be so designed that any deformation due to linear acceleration or other angular accelerations will either provide a minimum deformation response or the individual gauge outputs will self cancel leading to an insignificant out

APPLICATIONS Extremely high values of angular acceleration are available for machine or turbine monitoring, although slip rings or other transmission systems may be necessary. Seek advice on mounting methods

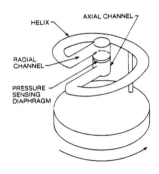

Representation of an angular accelerometer
Courtesy Endevco (UK)

RANGE LIMITS ±5000 radians/s^2 to $\pm50\ 000$ radians/s^2. 1 Hz to 600 Hz.	
ERROR Nonlinearity and hysteresis: $\pm1\%$ of full range.	
OVER-RANGE PROTECTION Transverse: 500 000 radians/s^2. Linear: Up to 2500 g.	
SUPPLY VOLTAGE 10 V dc to 14 V dc.[1]	
SENSITIVITY In the order of 20 mV/V of input supply.	
CROSS SENSITIVITY Transverse angular sensitivity: $\pm1\%$ of full range for all axes.[2]	
TEMPERATURE RANGE Operating: $-18°C$ to $+120°C$.[3]	
TEMPERATURE SENSITIVITY Range: $\pm0.03\%$ of full range/°C over compensated range.[4]	

SIGNAL CONDITIONING Typically suitable dc amplifiers and power supplies are required. Conventional strain gauge amplifiers are acceptable subject to suitability and compatibility.

BENEFITS AND LIMITATIONS For large angular acceleration values there is no competitor.

COMMENTS An interesting design problem to ensure a transducer with good angular response with minimal linear response.
[1] Only the recommended voltage should be used since other voltages may result in a deterioration of performance.
[2] The linear cross sensitivity will be significantly lower.
[3] Normally the compensated range is limited to a 50°C span.
[4] These devices are not recommended for use below 1 Hz and thus "Zero" temperature effects are of little import.

SUPPLIERS END1; END2.

AA2. Servo or force balance type

OPERATING PRINCIPLE Similar in operation to the equivalent linear device (AC8), but here the seismic mass is balanced about its rotational position. Insensitive to linear acceleration, the angular acceleration results in displacement of the mass which is sensed by a position sensor which produces an electrical signal used to provide a negative feedback – one which restores the balanced mass to its original position. This signal is a measure of the angular acceleration.

APPLICATIONS Usually fitted by bolting into position and because of their high accuracy are used for measurement of acceleration on aircraft, ships, helicopters and similar mobile applications.

Courtesy Lucas Schaevitz Ltd

RANGE LIMITS ± 200 radians/s^2 to ± 1000 radians/s^2. 50 Hz to 120 Hz dependent upon range.

OVER-RANGE PROTECTION Acceleration typically up to 100 g in all three axes.

ERROR Nonlinearity: 1 Radian/s^2. Hysteresis: 0.02% of full range.

SUPPLY VOLTAGE Most low voltage dc power inputs.

SENSITIVITY ± 5 V for full range.

CROSS SENSITIVITY Between $\pm 0.1\%$ and $\pm 0.2\%$ of full range.

TEMPERATURE RANGE Operating: $-55°C$ to $+95°C$. Storage: $-65°C$ to $+105°C$

TEMPERATURE SENSITIVITY Zero: 0.002% of full range/°C. Range: 0.025% of full range/°C.

SIGNAL CONDITIONING As described above, the signal conditioning is incorporated within the transducer. With the high dc output, further amplification is seldom required.

BENEFITS AND LIMITATIONS Closed loop systems allow the natural frequency and damping to be controlled electrically, permitting much greater control over the design parameters.

COMMENTS Like their linear counterpart these devices are likely to be the most precise way of measuring angular acceleration. The linear and angular devices are very similar in operation, the essential difference being only in the way in which the "pendulum" mass is either offset (linear) or balanced (angular).

SUPPLIERS LUC1; LUC2.

Angle

Although the measurement of angle (angular position) is a less ancient skill than its linear counterpart, there is no doubting its antiquity. Other than the well-known association of Pythagoras and Euclid with geometry, the need to establish a horizontal building base, zero angle, was well established by the time the Great Pyramid was built. Measurement of small angles was a vital part of early astronomy and the related science of navigation with the use of the quadrant, cross staff and astrolabe. In general terms, the measurement of angle, for most normal engineering needs, is by one of two techniques.

It may be determined by comparison with a vertical, the vertical plane in turn being determined by the Earth's local gravitational pull. It is for this reason that accelerometers and force measurement devices can be and are used for angular measurement. (Some are included here but the qualities are identified for devices which are purpose-built for this task rather than the general run of transducers described elsewhere in this book.)

Alternatively, angle is measured by comparison between the transducer housing and some form of shaft-driven rotor. This provides a measure of angular relationship between two positions, one to the other, independent of their general angular attitude. These devices are often supplied by the manufacturers in standard synchro sizes.

They may be used for establishing a local velocity by incorporating a measurement of time but where their measurement range is limited to 360° or less their main function is likely to be for the measurement of angle alone.

A further extremely common measurement of angle is achieved with a **navigational compass** – either magnetic or gyroscopic.

The conventional magnetic compass, using a freely-pivoted magnet, is of great antiquity but more recently a device known as the **flux-gate compass** has been introduced to provide a solid-state device coupled with a convenient read-out with controllable damping.

The **gyroscopic** compass too provides a similar service to its magnetic counterpart and it is not always appreciated that their use is wide-spread and not just limited to ship or aircraft navigation. Gyroscopes have the additional potential of measuring in the three mutually perpendicular planes from a single device.

The accurate measurement of angle is somewhat different to its linear counterpart in that the maximum range of 360° is fixed. For a device measuring through the full 360° with an accuracy of 0.1% the precision of measurement is 0.36° – an angle which the human eye can detect readily which in many applications would be too inaccurate.

A mechanical **clinometer** capable of resolving to 1 minute of arc and with a 360° capability requires 21,600 scale divisions or equivalent. This compares to a metre rule having scale markings at every 0.0463 mm and yet this quality of measurement is readily achievable.

In spite of the competition, electrical devices exist which offer a variety of capabilities and quality of measurement which will often make them preferable to mechanical techniques as well as being more "user-unfriendly". In the hands of staff capable of reading a vernier gauge or micrometer a clinometer offers no difficulty. However for the untrained user or where usage is not on a regular basis the electrical read-out offers considerable benefit.

Again where the angle is only a means to deriving some other parameter the ability to electrically manipulate the output without human interfacing is of considerable advantage as is the capability of some devices to be "zeroed". This allows the measurement of angular change without the complication of taking two readings and then subtracting to calculate that change.

Calibration of clinometers may be achieved by placing them on some form of table which can be set to known angles. One simple way is to use a dividing head, a device common in most modern workshops. Here a flat plate, holding a clinometer, is mounted in the dividing head's jaws and then a wide range of angles can be set. By using the dividing head chuck directly the angular position devices may also be calibrated in a similar manner.

The confirmation of the "zero" value of a clinometer is possibly the easiest to check of all transducers. This is achieved simply by placing it on a nominally horizontal surface, taking a reading and then rotating the device through 180° about its vertical axis and then repeating the reading. The algebraic average will be the zero value.

Some angle sensors or clinometers are identified as tilt sensors. They are still gravity-sensitive and it would appear that the term clinometer may be applied to devices which are self-contained complete with visual read-out, in line with the mechanical counterparts, whereas tilt sensors refer to the sensor alone. Whatever the subtlety "clinometer" will be used here for gravity-sensitive devices with "angular position" for the others.

Occasionally clinometers may be described as **inclinometers.** However since the latter is used specifically for the measurement of the vertical intensity of a magnetic force, a dipping compass, clinometer is preferred in the text.

Tilt sensors should not be confused with tilt switches which operate at some predetermined angle alone. Two such devices are included in the body of this Section.

NAMAS accredited laboratories: 0020, 0026, 0036, 0063, 0067, 0089, 0105, 0108, 0180, 0233, 0239.

All of the above laboratories provide clinometer calibration services. There is a very wide choice of laboratories who provide services under the NAMAS headings of angle gauges and angle plates.

Contents: Angle

AN1. Angular position, capacitive
0 – 8° to 0 – 270°.

AN2. Angular position, compass, flux-gate magnetometer
0 to 360°.

AN3. Angular position, gyroscopic
Yaw (azimuth): 360° continuous. Pitch (altitude): ±85°. Roll: ±360°.

AN4. Angular position, potentiometric
0 – 360°.

AN5. Angular position, rotary variable differential transformer (RVDT)
±30° to ±60°.

AN6. Angular position, transformer type
±2° to ±150°.

AN7. Clinometer, capacitive
±10° to ±90°.

AN8. Clinometer, electrolytic
±0.5° to ±80°.

AN9. Clinometer, inductive
±6° to ±30°.

AN10. Clinometer, potentiometric
±45° to ±90°.

AN11. Clinometer, servo system
±1° to ±90°.

AN12. Clinometer, transformer
±2° to ±150°.

AN13. Tilt switch, electrolytic
Depends upon setting.

AN14. Tilt switch, mercury
±13° to ±45°.

Angle

AN1. Angular position, capacitive

OPERATING PRINCIPLE Although dependent upon individual design, a typical example comprises capacitor plates, one attached to a rotating spindle, and two intertwined or interlinked stator plates combined to form a complete differential capacitor.

Thus as the rotor is moved the capacitance of one decreases as the other increases.

This differential capacitance change, in the form of a bridge circuit, is directly proportional to the angle of rotation.

APPLICATIONS A wide variety of uses such as for control linkages, providing both position indication and feedback signal.

Supported through a bracket mounting and drive shaft.

Courtesy Camille Bauer Controls Ltd

RANGE LIMITS 0 – 8° to 0 – 270° in a number of ranges.

ERROR Nonlinearity: ±0.01% to ±0.5% of full range.

OVER-RANGE PROTECTION Continuous rotation or mechanical stops fitted.[1]

SUPPLY VOLTAGE 12 to 36 V dc or mains power.

SENSITIVITY Resolutions of 1 arc second when complete with signal conditioning.

CROSS SENSITIVITY Insignificant.

TEMPERATURE RANGE Operating: –25°C to +70°C. Storage: –40°C to +70°C

TEMPERATURE SENSITIVITY ±0.03% of full range/°C (ambient conditions).[2] ±0.05% of full range/°C (complete range).

SIGNAL CONDITIONING Usually supplied within the transducer. There are a number of options including zero and range adjustment with voltage or current loop output.

BENEFITS AND LIMITATIONS They have low rotational torque and can be used to convert simple angular motion into linear control signals.

COMMENTS Because of the low driving torque it is possible to use magnetic coupling when a liquid-tight seal is required.
[1] Although reasonably immune to cross sensitivity effects, care should be taken to prevent damage caused by over rotation against the mechanical stops, or by the application of excessive side or bending forces to the drive spindle.
[2] Quoted for 0.5% linearity devices.

SUPPLIERS AUT; CAM1; CAM2.

Angle

AN2. Angular position, compass, flux-gate magnetometer

OPERATING PRINCIPLE Two coils placed at right angles to each other detect the Earth's magnetic field.

Since the resultant field strength will depend upon the attitude of each coil to the magnetic lines of flux, through comparison, one to the other, it is possible to derive from the resultant of the coils the angular direction of the Magnetic North Pole.

The information contained here is specifically for devices sold for marine use.

APPLICATIONS Usually fitted to smaller ships and pleasure craft.

Great care must be taken to install the compass free from extraneous magnetic fields.

Courtesy NASA (1992) Ltd

RANGE LIMITS 0 to 360°, continuous rotation.

ERROR Accuracy: ±1° of angle. (±0.3% of full 360°).

OVER-RANGE PROTECTION Not applicable if properly installed.

SUPPLY VOLTAGE 12 V and 24 V dc or mains power.

SENSITIVITY ±1° or better, subject to display.

CROSS SENSITIVITY The Earth's magnetic field is a three dimensional field. Thus as the ship pitches or rolls the compass will experience a change of flux even for a constant compass bearing. These devices include electronic correction for such potential sources of error.

TEMPERATURE RANGE Operating: −25°C to +70°C. Storage: −30°C to +70°C.

TEMPERATURE SENSITIVITY Essentially free of temperature effects.

SIGNAL CONDITIONING Offers a complete system including display and electrical output. Flux-gate compasses often form part of another system such as a ship's autopilot.

BENEFITS AND LIMITATIONS These devices offer much over the conventional magnetic compass, however they are prone to the same sources of error.

COMMENTS Flux-gate compasses, in keeping with their traditional counterparts, are affected by local magnetic interference known as deviation. Correction may be necessary. Since "true" North is required the compass may incorporate a variation correction. Ship motion through the water is such that there are continuous small changes of heading and the smaller the vessel the more frequent and violent is this angular change. Damping is required to smooth this undesired effect. Controllable electronic damping is normally incorporated.

SUPPLIERS NAS; TEL.

AN3. Angular position, gyroscopic

OPERATING PRINCIPLE The gyroscope is a rotating disc mounted on a pair of gimbal mounts known as the inner and outer gimbals.

The rotating flywheel maintains a constant position as the gimbal mounts are moved.

A measure of these relationships is the angular position.

Gyroscopes are "free" when the angle is referred to the casing on spin-up; or "directional" where the altitude and heading are determined by external references.[1]

APPLICATIONS Used as a compass on large ships.

Other uses include aerospace and a wide range of other mobile applications and research.

Manufacturer's operating instructions must be followed meticulously.

Courtesy Humphrey Inc

RANGE LIMITS Yaw (azimuth): 360° continuous.[2] Pitch (altitude): ±85°.[2,3] Roll: ±360°.

ERROR Nonlinearity: ±1.0% of full range. Accuracy: 15 arc minutes or better[4] to ±0.5° subject to type.[5]

OVER-RANGE PROTECTION Shock up to 60 g. Vibration levels up to 15 g.

SUPPLY VOLTAGE Gyroscopes are ac or dc driven such as 26 or 115 V, 400 Hz ac or 28 V dc.

SENSITIVITY Dependent upon type of pick-off and display.

CROSS SENSITIVITY Transverse sensitivity is incorporated within the error values.

TEMPERATURE RANGE All normal environmental temperatures or greater.

TEMPERATURE SENSITIVITY Insignificant except for possible errors arising from the signal pick-off.

SIGNAL CONDITIONING These systems come complete with all necessary electronics and caging facilities.

BENEFITS AND LIMITATIONS A unique device since it operates by relating its attitude to its own reference system; called an inertial frame of reference.

COMMENTS [1]For navigation, an azimuth reference is required such as a magnetic flux-gate device to maintain a correction as the gyroscope moves around the Earth's surface. These are called "slaved, directional". [2]Identified here in both aeronautical and (in parenthesis) navigational terms. [3]The rotation between the inner and outer gimbal is limited to below ±90° to protect the gyroscope from an effect known as "gimbal lock". [4]Synchro pick-offs. [5]Potentiometric pick-offs.

SUPPLIERS HUM; IFO; SMI.

AN4. Angular position, potentiometric

OPERATING PRINCIPLE A resistance track is housed in the body of a rotary potentiometer and a rotating shaft guides a wiper across this resistance element. A voltage is supplied across the resistor so that a measurement of the voltage between the slider and either end of the resistance element provides a measure of displacement. The resistance track is either wirewound or a hybrid of conductive plastic and wire, the latter providing a much better resolution.[1]

APPLICATIONS Very wide-ranging including linkage and control rod systems such as robot arms.
Mounting is simple by suitable brackets and proper shaft alignment attachment.

Courtesy Penny & Giles Ltd

RANGE LIMITS Up to 0 to 360°, continuous rotation.[2]

ERROR Nonlinearity: ±0.05% to ±0.15% subject to angular range.

OVER-RANGE PROTECTION Since the devices can be continuous rotation or fitted with angle stops, protection may be judged as "high".

SUPPLY VOLTAGE Up to 130 V dependent upon the power dissipation of the sensor.

SENSITIVITY Full range voltage is coincident with the supply voltage.

CROSS SENSITIVITY Insignificant. Although these devices use a rubbing electrical contact "life" values of 100×10^6 cycles are quoted.

TEMPERATURE RANGE Operating: –55°C to +90°C.

TEMPERATURE SENSITIVITY Resistance change is in the order of 120 ppm/°C.

SIGNAL CONDITIONING The voltage available at full scale is a large dc potential thus any further signal conditioning or amplification is unlikely.[3]

BENEFITS AND LIMITATIONS Rugged and simple transducers. Although these devices use a sliding contact the long-term performance is good.

COMMENTS The input voltage may be set to provide an output reading directly in engineering units.
[1]The hybrid tracks should be used only in a potentiometric voltage divider mode whereas the wire-only versions may be used in the simple resistance mode.
[2]There may be some signal deterioration at the 0 to 360° position. This may limit the total range, such as 0 to 358°.
[3]Hybrid devices must be used as voltage dividers and hence require a high output impedance.

SUPPLIERS PEN; PRE; RSC.

Angle

AN5. Angular position, rotary variable differential transformer (RVDT)

OPERATING PRINCIPLE An RVDT operates in a similar manner to the LVDT displacement device and is constituted in the form of a primary excitation winding and two secondary windings mounted in a casing.

The arrangement of these coils is such that the angular displacement of a rotor-driven core affects the transformer coupling, primary to secondaries, and produces a differential output voltage directly proportional to the rotor angular position.

APPLICATIONS Most suitable where long-term stability is required. These include military, aerospace and control systems.

Mounting is simple by suitable brackets and proper shaft alignment attachment.

Courtesy Lucas Schaevitz Ltd

RANGE LIMITS ±30° to ±60°.[1]
ERROR Nonlinearity: ±0.25% to 1.5% of full range; ±0.1% for small angles.[2]
OVER-RANGE PROTECTION The shaft is free to rotate continuously. However care must be taken to protect the drive shaft from excessive side force.
SUPPLY VOLTAGE 3 V(rms) at a frequency of between 400 Hz and 20 kHz.[3]
SENSITIVITY Typically 125 mV/degree.
CROSS SENSITIVITY Insignificant.
TEMPERATURE RANGE Operating: –50°C to + 150°C.
TEMPERATURE SENSITIVITY Temperature coefficient: 0.04% of full range/°C.
SIGNAL CONDITIONING Dependent upon the requirement. Some transducers incorporate signal conditioning. This is then energised by low voltage dc.
BENEFITS AND LIMITATIONS An extremely stable measuring system with very low torque required to drive the shaft. The sinusoidal output can be advantageous.
COMMENTS [1]These devices produce a sinusoidal output of angle. Hence these angles are for the best-fit linear part of a sine curve. [2]These changes in nonlinearity arise from fitting a best-fit straight line through a sinusoidal output. [3]400Hz is a common aircraft supply frequency and thus direct use of this supply can be incorporated into the measuring system.
SUPPLIERS LUC1; LUC2.

Angle

AN6. Angular position, transformer type

OPERATING PRINCIPLE These devices are essentially a transformer in which the coupling, and hence the output from the secondary winding, is modified by the angular position of the input shaft in relation to the body of the transducer. Like all transformer-type devices, there is electrical isolation between the input supply and output signal. These devices must not be confused with the RVDT transducer as described at AN5.

APPLICATIONS A wide range of engineering and industrial uses with a low torque drive requirement.
Attachment by suitable brackets and shaft drive.

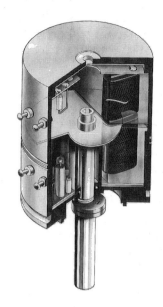

Courtesy Penny & Giles Ltd

RANGE LIMITS ±2° to ±150°.

ERROR Nonlinearity: ±0.5% of full range.

OVER-RANGE PROTECTION The shaft is free to rotate continuously. However care must always be taken to protect the drive shaft from excessive side forces.

SUPPLY VOLTAGE 10 V dc to inbuilt signal conditioning.

SENSITIVITY Between 470 mV/degree and 33 mV/degree of full range.

CROSS SENSITIVITY Insignificant.

TEMPERATURE RANGE Operating: –20°C to +60°C. Storage: –40°C to +100°C.

TEMPERATURE SENSITIVITY ±0.8 mV/°C between +10°C and +40°C.

SIGNAL CONDITIONING Incorporated within the transducer casing.

BENEFITS AND LIMITATIONS A self-contained device, particularly suitable for remote operation. It can be supplied for "current loop" operation.

COMMENTS A device which offers convenient measurement without remote signal conditioning being required. Although its performance is likely to be less precise than the RVDT, it will have the same long-term stability associated with inductive or transformer devices.

SUPPLIERS PEN.

AN7. Clinometer, capacitive

OPERATING PRINCIPLE A liquid is contained within a circular chamber of which the two end surfaces incorporate capacitor plates; the liquid forms the dielectric of the capacitance.

As the device is tilted the liquid settles to a new position causing a change in capacitance.

Additionally a two-axis device is manufactured which has the appearance of a bubble clinometer in which the movement of the bubble position identifies the two tilt angles.

APPLICATIONS Typical easy-to-use clinometer which may form a permanent part of an installation or a truly portable clinometer.
Permanent attachment is by bolts.

Courtesy Lucas Schaevitz Ltd

RANGE LIMITS ±10° to ±90° in a limited number of ranges.[1]

ERROR Nonlinearity, 0 to 20°: ±0.05°. 20 to 60°: ±0.1°. Other angles: ±1% of reading. Repeatability: ±0.05°.

OVER-RANGE PROTECTION None required. Violent motion may result in erroneous readings but the device will return to proper performance after being allowed to settle.

SUPPLY VOLTAGE Low voltage dc for remote devices, or with fitted batteries for self-contained versions.

SENSITIVITY 16 to 52 mV/degree (ratiometric mode); 60 mV/degree (analogue mode); ±0.1° for visual display.

CROSS SENSITIVITY Cross axis tilt: less than 1% of full range for 45° tilt.

TEMPERATURE RANGE Operating: –40°C to +65°C. Storage: –55°C to +65°C.

TEMPERATURE SENSITIVITY Zero: ±0.008°/°C. Range: ±0.05% of full range/°C

SIGNAL CONDITIONING Complete with the sensor. Ratiometric, analogue and pulse width modulation systems are offered for remote units and visual display for self-contained versions.

BENEFITS AND LIMITATIONS A reliable system with no mechanical moving parts which offers a very long life. These self-contained devices are extremely easy to use.

COMMENTS Perhaps the nearest electrical equivalent to the mechanical bubble device. Its time constant, up to 2 s, prohibits its use for dynamic applications. Available both as remote and self-contained clinometers. Outputs in various electrical forms for the former or a visual display for the latter which have optional display forms. [1]Dependent upon the style of casing, devices which accommodate ±45° can be used for angles up to 360° by use of different surfaces of the casing.

SUPPLIERS LUC1; LUC2.

Angle

AN8. Clinometer, electrolytic

OPERATING PRINCIPLE A chamber, partially filled with an electrolyte, has two electrodes or probes inserted such that their free ends penetrate through the liquid into the air space. A third electrode is also inserted into the liquid but remains fully immersed at all angles of tilt.

Tilting changes the depth of immersion of the partially-immersed probes and a measure of the resistance change between them and the third probe is a direct measure of angle.[1]

APPLICATIONS Military, offshore structures, ships, mines, and is well suited for use in harsh physical environments.

Mounting is generally simple by bolting down.

Courtesy Tilt Measurement Ltd

RANGE LIMITS ±0.5° to ±80°.[2]

ERROR Nonlinearity: ±0.1% to ±5% of full range.[3] Repeatability: ±0.05% to ±1% of full range.

OVER-RANGE PROTECTION High. Shock resistant up to 300 g. After excessive motion, on return the device will quickly settle back to full operation to steady conditions.

SUPPLY VOLTAGE 4 to 12.5 V ac, 50 Hz to 20 kHz.

SENSITIVITY Between 5 and 100 mV/V supply/°.

CROSS SENSITIVITY 0.7% of full range/° of transverse tilt is an example. However transverse sensitivity is dependent upon the transducer.

TEMPERATURE RANGE Operating: −71°C to +100°C.

TEMPERATURE SENSITIVITY Zero: 0.05% of full range/°C. Range: 0.1% of full range/°C

SIGNAL CONDITIONING A minimum requirement is a stable ac supply, a bridge completion network and an ac rectifier. Integrated systems, dc-energised, are available.

BENEFITS AND LIMITATIONS A very simple and essentially robust device. For the larger ranges some transducers are not linear. Two-axis devices are available.

COMMENTS A neat, simple and robust solution for an extremely wide range of applications. [1]These devices come in essentially three shapes: cup shaped, circular arc similar to a bubble spirit level, and toroidal, similar to the circular arc version but encompassing a greater angular range.
[2]Some devices are quoted for both a linear range and a much larger unambiguous but nonlinear one.   [3]Limited to between ±0.5° and ±20°.

SUPPLIERS IFO; THE; TIL.

AN9. Clinometer, inductive

OPERATING PRINCIPLE Based upon a pendulum whose position is detected by two inductive coils.

When at zero angle, a ferromagnetic ball, the pendulum "bob", lies equidistant between the two coils.

As the device is tilted there is a corresponding positional movement of the bob causing appropriate changes in inductance between the two coils. This change, converted into a dc signal, is proportional to the tilt angle. The whole system is suitably damped.

APPLICATIONS Simple to mount by bolting, this device is robust and reliable. Hence it is widely used in marine, ground movement, industrial and similar applications.

Courtesy RDP Group

RANGE LIMITS ±6° to ±30°	
OVER-RANGE PROTECTION Generally robust.	
ERROR Accuracy: ±0.25% of full range to ±0.5% of full range.	
SUPPLY VOLTAGE 12 to 30 V dc, energising self-contained signal conditioning.	
SENSITIVITY ±1 V dc output for full range from signal conditioning.	
CROSS SENSITIVITY Insensitive to cross axis tilt error up to 50°. However it is advisable to limit tilting to ±15°.	
TEMPERATURE RANGE Compensated: –1°C to +49°C. Operating –23°C to +80°C. Storage –55°C to +100°C.	
TEMPERATURE SENSITIVITY Zero ±0.04% of full range/°C. Range ±0.04% of full range/°C.	
SIGNAL CONDITIONING The signal conditioning is contained within the transducer housing. Thus further amplification may not be necessary.	
BENEFITS AND LIMITATIONS It has a high natural frequency for a pendulous device, which for some ranges can be in excess of 10 Hz.	
COMMENTS A straightforward application of electrical technology to replace the mechanical equivalent. Being self-contained with a good signal output it lends itself for incorporation into multichannel measurement packages.	
SUPPLIERS RDP.	

AN10. Clinometer, potentiometric

OPERATING PRINCIPLE A pendulum is suspended from the shaft of a rotary potentiometer. Thus rotation of the body of the transducer results in the wiper contact of the potentiometer being moved across the resistance element.

The resulting voltage output is directly proportional to the angular position.

APPLICATIONS Being simple, rugged devices, they may be used in a wide range of applications where the resolution (10 arc minutes) and overall accuracy are sufficient for the chosen task.

Mounting is simple and achieved by bolting down.

Although these devices use a sliding contact, they have a long operating life, typically 20×10^6 cycles.

Courtesy Penny & Giles Ltd

RANGE LIMITS $\pm 45°$ to $\pm 90°$.

ERROR Nonlinearity: $\pm 0.25\%$ of full range. Hysteresis: $\pm 1°$ or better.

OVER-RANGE PROTECTION High. One manufacturer quotes up to a maximum shock in any axis of 100 g.

SUPPLY VOLTAGE Up to a maximum of 35 V dc.[1]

SENSITIVITY Full range voltage is coincident with the supply voltage.

CROSS SENSITIVITY Insignificant.

TEMPERATURE RANGE Operating: –50°C to + 100°C

TEMPERATURE SENSITIVITY Temperature coefficient of resistance: 120×10^{-6} $\Omega/\Omega/°C$.

SIGNAL CONDITIONING The voltage which is available at full scale is so large it is unlikely that further amplification will be required.[2]

BENEFITS AND LIMITATIONS A simple rugged device which requires no amplification.

COMMENTS By manipulating the input voltage to some predetermined value the output voltage could read directly in "angle" or any other desirable quantity such as "grads".

[1]The centre position (0°) will be at the mid-voltage value of the supply.

[2]Care should be taken to ensure that high currents are not drawn through the wiper contact or the potentiometer winding. It is good practice to use the device as a voltage divider into a high impedance source.

SUPPLIERS PEN

Angle

AN11. Clinometer, servo system

OPERATING PRINCIPLE A mass, simulating the "bob" of a pendulum, senses the applied acceleration due to local gravity.

The resulting displacement is sensed by a position sensor which produces an electrical signal which is used as negative feedback to provide both a corrective moment to the pendulum arm, restoring the mass to its original position, and a measure of the angle of tilt.[1]

APPLICATIONS Used for weapons, platforms, civil engineering levelling and grading, crane tilting and indeed for many precision angular measurements, particularly when used in a control system.[2]

Courtesy Lucas Schaevitz Ltd

RANGE LIMITS ±1° to ±90°.

ERROR Nonlinearity: ±0.002% of full range. Repeatability and hysteresis: from ±0.02% of full range to ±0.0005% of full range as the angular range increases.

OVER-RANGE PROTECTION Acceleration: 50 g with shock loading considerably higher.

SUPPLY VOLTAGE Low voltage dc such as a nominal 9 V supply.

SENSITIVITY ±5 V dc full range, proportional to the sine of the tilt angle.

CROSS SENSITIVITY Cross axis tilt: better than ±0.001 g/g.

TEMPERATURE RANGE Operating: –18°C to +71°C. Storage: –40°C to +75°C.

TEMPERATURE SENSITIVITY Zero: ±0.05% of full range/°C reducing to ±0.003% of full range/°C as the angular range increases. Range: ±0.04% of full range/°C reducing to 0.006% of full range/°C as the angular range increases.

SIGNAL CONDITIONING Supplied complete with all signal conditioning to provide a dc output proportional to the sine of the angle.

BENEFITS AND LIMITATIONS These devices have high accuracy. As a replacement for mechanical clinometers they are limited in angular range but offer remote operation.

COMMENTS [1]With this type of device there is further data manipulation required to provide the angular value. However there are many occasions where the "sine" output is preferred when used for position control.
[2]They are available both as remote transducers and complete with signal conditioning and visual display. Damping is achieved electrically.

SUPPLIERS LUC1; LUC2.

AN12. Clinometer, transformer

OPERATING PRINCIPLE The design is based upon a pendulum whose relative angular position moves a transformer core which in turn changes the coupling, primary to secondary, of the transformer windings.
The resulting electrical change, converted into a dc signal, is proportional to the tilt angle. The whole system is suitably damped by mechanical means.

APPLICATIONS Simply mounted with remote display, the system is used on ships, offshore structures, crane jibs and where a reliable robust system is required.

Courtesy Penny & Giles Ltd

RANGE LIMITS $\pm2°$ to $\pm150°$.

ERROR Nonlinearity : ±0.4 minutes of arc to ±0.25 minutes of arc.[1] Hysteresis: 1 minute of arc for high accuracy devices.[2]

OVER-RANGE PROTECTION Generally robust and may be rotated through 360° continuous rotation. Up to 400 g vertical shock.

SUPPLY VOLTAGE With signal conditioning 10 V dc is required.

SENSITIVITY Full range output up to 10 V dc.

CROSS SENSITIVITY Claimed to be insignificant to cross-axis angular position.

TEMPERATURE RANGE Operating: $-20°C$ to $+60°C$ or better. Storage: $-40°C$ to $+100°C$.

TEMPERATURE SENSITIVITY ±0.8 mV/°C over $+10$ to $+40°C$ for general-purpose devices.[3]

SIGNAL CONDITIONING Self-contained within the transducer housing requiring a dc input and with a dc output.

BENEFITS AND LIMITATIONS These devices can be obtained in two-axis units. In addition to the shock resistant version others include sealing for water immersion and industrially rugged versions.

COMMENTS Similar in operation to the angular position version described at AN6 to which it may be compared.
[1] Including temperature effects.
[2] For angles between ±2 and $\pm30°$.
[3] Increasing to ±2 mV/°C outside this band.

SUPPLIERS PEN.

AN13. Tilt switch, electrolytic

OPERATING PRINCIPLE An electrolytic liquid partially fills a capsule, forming a gas bubble at the top.

Two electrodes are inserted into the liquid, one of which passes into the bubble.

By insulating this probe except for its tip, an electrical current will only flow between these electrodes as tilting takes place when both electrodes are immersed in the electrolyte.

This switch change may then be used to operate some control mechanism or device.

APPLICATIONS Used for control and alarm applications over a wide range of applications for precision detection and/or where the use of mercury is prohibited.

Attachment by bolting down.

Courtesy IFO International (UK) Ltd

RANGE LIMITS Preset to the desired angle.	
ERROR Hysteresis: ±15 minutes of arc.	
OVER-RANGE PROTECTION No limit to tilt angle.	
SUPPLY VOLTAGE Low voltage ac is required and values up to 26 V are quoted.	
SENSITIVITY Discrimination as high as 0.1 arc minute.	
CROSS SENSITIVITY Transverse sensitivity values are low. (See also AN8)	
TEMPERATURE RANGE –55°C to +100°C.	
TEMPERATURE SENSITIVITY Insignificant.	
SIGNAL CONDITIONING Unregulated ac supply is the minimum but simple rectification will be needed for the operation of dc switching devices.	
BENEFITS AND LIMITATIONS For tilt in the same plane but in opposite angular sign, two devices are required since they operate in one direction only.	
COMMENTS A high precision device with controllable damping. It is more accurate than the mercury device (AN14) and is safer to use. Its use as an ac or dc output will depend upon both requirements of the switching and of the availability of a supply.	
SUPPLIERS IFO; THE; TIL.	

AN14. Tilt switch, mercury

OPERATING PRINCIPLE Two electrical contacts are in close proximity to a small bead of mercury, all contained within a sealed encapsulation.

When the switch is tilted, the mercury wets the two contacts and this ensures electrical continuity between them.

These devices have a long history of use and their freedom from contact "bounce" has ensured their continued use for modern logic circuitry.

APPLICATIONS Many applications particularly in the security or safety fields.

They allow ready conversion of existing mechanical devices to ones with an electrical control such as for float switches.

Courtesy RS Components Ltd

RANGE LIMITS ±13° to ±45°.[1]

ERROR A hysteresis value of 15° is an example, depending upon the design.

OVER-RANGE PROTECTION None required. Devices will accommodate considerable increase in angle, such as 170°, whilst maintaining the switched mode.

SUPPLY VOLTAGE Limited by the resistance of the contact, with values as low as 50 mΩ.[2]

SENSITIVITY Angular position sensitivity: better than 1°.

CROSS SENSITIVITY Influenced by the design which may allow only one plane of tilting. However in general use the transverse sensitivity is low.

TEMPERATURE RANGE Not quoted but should meet most environmental requirements.[3]

TEMPERATURE SENSITIVITY Insignificant.[4]

SIGNAL CONDITIONING None required. As identified these devices are ideal for bounce-free switching applications.

BENEFITS AND LIMITATIONS A simple, low cost and robust device which has a long established record of reliable performance.

COMMENTS [1]In various arrangements, with both contacts closed or open at their initial position.
[2] Such a resistance allows a switching current of up to 0.5 A for voltages from 28 V dc to 240 V ac.
[3]The melting point of mercury is approximately –40°C.
[4] Mercury has a low coefficient of expansion: 60×10^{-6}/°C. This compares with aluminium the value for which is 23×10^{-6}/ °C.

SUPPLIERS RSC.

Flow

Where and when in human history the earliest interest in gas flow measurements arose is unknown but it is not unreasonable to suggest that the natural flows of winds with their effects upon the environment must have been of great import.

The effects of wind velocity must have been well known and possibly considered with fear and respect, certainly by the early mariners, but the first modern attempt to quantify wind speed was the Beaufort Scale devised by Admiral Sir Francis Beaufort in 1808. This was a technique whereby the sea motion was used as a "transducer" to indicate a measure of the wind velocity. Later, alternatives such as tree motion were added for "landlubbers".

Perhaps one of the most interesting equivalent scales is the need for flying creatures to stay grounded at some limiting wind speed – from midges to geese. These flying creatures perhaps are even better calibrated transducers than the sea.

For liquid flow, the beginnings of understanding will certainly have been associated with irrigation and early canals.

The nautical speed unit of the knot, or 1 nautical mile per hour, was the first formal method of identifying speed through water. The choice arises because the nautical mile is the length of one minute of arc along a meridian of longitude and hence had and continues to have practical navigational uses.

The measurement of fluid flow may be categorised into two simple divisions:

  a) Flow in pipes or ducts or their associated inlet or exhaust;
  b) Unrestrained flow, almost exclusively limited to air or water.

Traditionally liquid flow, either in pipes or open ducts, was achieved by measuring the pressure drop across some restriction such as an **orifice plate.** Indeed these techniques are very common. Where such pressure is measured directly, the quality of measurement is determined by the chosen pressure measuring device which are included in a separate section of this book.

The simplest of all of the portable devices is the **Pitot tube** (liquid) or Pitot static tube (gases) and this has a very high order of accuracy combined with the benefit of no moving parts. However as a sensor it does not have an electrical output without the use of an attached pressure sensor and the accuracy is dictated mainly by the quality of the chosen sensor.

NOTE. Some instruments are applicable for both gases and liquids. To save repetition these will be described under gases with a brief table identifying their location under liquids.

Flow, gases

The most commonly seen measurement transducers of gas flow are those associated with wind velocity. Known as **anemometers,** these devices are readily seen on ships, yachts, cranes and buildings in addition to their use for meteorological purposes.

In that latter role the cup anemometer is the reference device and their use for weather data collection is controlled by strict codes of practice to ensure a commonality in interpretation. Thus the preferred height is dictated as is the definition of mean wind speed and gusts.

This formal interpretation of wind velocity is not necessarily a precise or true measure of the local velocity but it is one which is supported by many decades of data. Thus, when comparing results from a cup anemometer with some other type of device, the relationship and interpretation will need careful consideration.

It is important to remember that for rotational devices, various optoelectronic counting techniques are used. However such methods do not qualify for inclusion here under the restrictions discussed at the beginning of the book.

Since some of the devices measure speed, as distinct from velocity, a further device is needed to measure direction. Such devices are included here rather than in the section dealing with angular measurement.

Included amongst the range of electrical techniques are ultrasonic and hot wire or film techniques which are capable of identifying the three dimensional local flow characteristics, but they are restricted in the main to where the flow experiences rapid variations such as for turbulence measurements. Their measurement quality and/or cost may prohibit their selection for steady flow measurement alone.

Normally the requirement for flow in pipes is to identify the volume or mass flow. Here flow is given in litres per minute (1 l/min $= 0.001$ m^3/min).

The calibration of velocity devices requires special wind tunnel equipment.

For the various rotor devices and wind vanes perhaps the more important aspect is to ensure proper and regular maintenance. It is advised that the manufacturer's spares should be used exclusively.

NAMAS Accredited Laboratories: 0009, 0072 and 0095.

Contents: Flow, gases

FG1. Pipe flow, mass flow, temperature
Gases: 5 ml/min to 17 000 l/min.
Liquids: 0.017 ml/min to 333 ml/min.

FG2. Pipe flow, variable area
Gases: 9 l/min to 60 000 l/min.
Liquids: 2 l/min to 1100 l/min.

FG3. Unrestrained flow, cup anemometer, pulsed output
0.25 m/s to 75 m/s.

FG4. Unrestrained flow, cup anemometer, self-generating
0.3 m/s to 50 m/s.

FG5. Unrestrained flow, hot wire/film anemometer
Gases: 15 m/s to 350 m/s.
Liquids: 0.03 to 30 m/s.

FG6. Unrestrained flow, rotating vane anemometer, capacitive
0.2 to 30 m/s.

FG7. Unrestrained flow, rotating vane anemometer, inductive
0.4 to 40 m/s.

FG8. Unrestrained flow, rotor, vertical wind speed
0.2 m/s to 50 m/s.

FG9. Unrestrained flow, thermal anemometer, thermistor
2 to 50 m/s.

FG10. Unrestrained flow, ultrasonic anemometer
Speed: 0 to 60 m/s. Direction: 0 to 360°, all axes.

FG11. Unrestrained flow, wind direction, potentiometer
0 to 360°.

FG12. Unrestrained flow, wind direction, reed switching
0 to 360°.

Flow, gases

FG1. Pipe flow, mass flow, temperature

OPERATING PRINCIPLES These devices operate by exploiting the characteristic of heat transfer under laminar flow conditions. To achieve the measurement, part of the main flow is diverted through a capillary tube which is heated from an external source. The heat absorbed by the gas or liquid is measured by establishing the temperature difference of the upstream and downstream ends of the capillary. This temperature differential is a direct measure of the mass flow.

APPLICATIONS Installed into the pipework generally by flange mounting, these devices are used wherever mass flow, as distinct from volume flow, is the preferred measurement

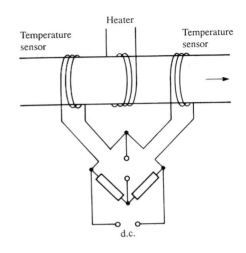

Thermal Mass Flowmeter Circuit
Courtesy Butterworth-Heinemann Ltd

RANGE LIMITS Gases: 0 – 5 ml/min to 0 – 17 000 l/min.[1] Liquids: 0 – 0.017 ml/min to 0 – 333 ml/min.[1]

ERROR Nonlinearity and hysteresis: ±1% of full range. Repeatability: ±0.2% of full range.

OVER-RANGE PROTECTION None required under normal circumstances. There is a limit to the ambient temperature of the fluid as specified below.

SUPPLY VOLTAGE Signal conditioning is generally incorporated hence typical mains voltages are used.

SENSITIVITY The output signal is typically 5 V for full range.[2]

CROSS SENSITIVITY Sensitive to pressure: ±0.001% of full range/kPa for gases. Insignificant for liquids.

TEMPERATURE RANGE Gases: –10°C to +70°C. Liquids: +10°C to +40°C.

TEMPERATURE SENSITIVITY Zero: ±0.05% of range/°C. (Over a 70°C range).[3] Range: ±0.1% of full range/°C.[3]

SIGNAL CONDITIONING Generally supplied as a complete system. The system can be supplied as a complete mass flow controller.

BENEFITS AND LIMITATIONS Mass flow measurement is essentially free of the effects of temperature and pressure.

COMMENTS A most interesting application of one of the properties of laminar flow.
[1] These ranges are for air and water respectively and will be different for other fluids.
[2] Other voltages and current loop outputs are available.
[3] These values are for both gases and liquids.

SUPPLIERS KRO.

FG2. Pipe flow, variable area

OPERATING PRINCIPLES Comprises a tapered glass tube within which is inserted a float. This forms an annular ring between pipe and float, the area of which changes as the float position alters.

As the fluid flows through the tube, the float begins to rise until the pressure differential across the float is sufficient to balance its weight.

Float position is detected by inductance or connected by levers to an angular capacitance transducer (AN1).

APPLICATIONS Generally connected in-line by conventional pipe fittings, these devices are excellent general-purpose flow transducers where remote or unambiguous readings are required.

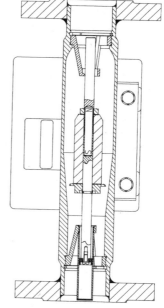

Courtesy Platon Instrumentation Ltd

RANGE LIMITS Gases: 0.5 – 9 l/min to 6000 – 60 000 l/min. Liquids: 0.2 – 2 l/min to 100 – 1100 l/min.

ERROR Accuracy: Between ±2% and 3% of full range. Repeatability: ±1% of full range.

OVER-RANGE PROTECTION None required. There is a pressure limitation and values of 4 MPa (gases) and 35 MPa (liquids) are quoted.

SUPPLY VOLTAGE Dependent upon type using both low voltage dc and normal mains.

SENSITIVITY Both low voltage dc and current loop systems are available.

CROSS SENSITIVITY These devices are sensitive to the Reynolds Number of the fluid.[1]

TEMPERATURE RANGE +5°C to +80°C.

TEMPERATURE SENSITIVITY Sensitive to the change of viscosity of the fluid being monitored.[1]

SIGNAL CONDITIONING Can vary from a simple analogue scale and pointer to a more complex data logger.

BENEFITS AND LIMITATIONS Perhaps the most common of all pipe flow meters in its mechanical form. An electrical output increases its versatility.

COMMENTS These devices are suitable for a very wide range of fluids and with careful selection have good chemical resistance.
[1] This non-dimensional unit is defined as the ratio between a fluid's inertial forces and its drag forces. The Reynolds Number is influenced by the flow rate, the specific gravity, the flow meter pipe dimensions and the viscosity of the fluid. Thus each device is calibrated for a given (supplied) fluid.

SUPPLIERS KRO; PLA.

Flow, gases

FG3. Unrestrained flow, cup anemometer, pulsed output

OPERATING PRINCIPLES A three arm "spider" has a cup of conical or hemispherical shape mounted at the end of each arm.
The mechanical centre is mounted upon a shaft using low friction bearings.
When located in a flow of wind the spider rotates around the vertically-placed shaft. Incorporated within the device is a switch such as a magnet and reed switch.
Each rotation produces a pulse which is counted over a given time interval.

APPLICATIONS Usually mast-mounted, the major use of these devices is to provide wind speed data or to give warning of potential disaster such as the overturning of cranes in high winds.

Courtesy Vector Instruments

RANGE LIMITS 0.25 m/s to 75 m/s in a single instrument.[1]

ERROR Accuracy ±1% of full range, ±0.1 m/s.[2]

OVER-RANGE PROTECTION In excess of 75 m/s. They will survive almost any wind condition, subject to the reliability of the supporting structure.

SUPPLY VOLTAGE Switching voltages of up to 100 V dc can be accommodated.[3]

SENSITIVITY One pulse/revolution is the minimum measurement resolution.

CROSS SENSITIVITY Cup anemometers respond to air flows perpendicular to the plane of rotation. Do not use where a vertical component of flow is present.

TEMPERATURE RANGE Operating: –30°C to +55°C or better. Storage: –40°C to +65°C.

TEMPERATURE SENSITIVITY Insignificant effects.

SIGNAL CONDITIONING Unless a complex programme is envisaged, it is advisable to use the manufacturers system usually available with a range of outputs.

BENEFITS AND LIMITATIONS Not suitable for serious fluctuating flow measurement because of the relatively high inertia causing slow response to rapid changes in velocity.

COMMENTS They have a long history of development and are reliable devices.
[1]Although essentially linear, there is a small bearing friction which requires an initial wind flow to start rotation often called the "breakout" or "threshold" value.
[2]Because of inertia effects, under wind gusting conditions, different styles of anemometers may well give different performance results.
[3]This is determined by the preferred switching voltage but will be dependent upon both the switch capability and the method of pulse counting chosen.

SUPPLIERS VEC.

Flow, gases

FG4. Unrestrained flow, cup anemometer, self-generating

OPERATING PRINCIPLES A three or four arm "spider" has a cup of conical or hemispherical shape mounted at the end of each arm.

The mechanical centre is mounted upon a shaft using low friction bearings.

When located in a flow of wind the spider rotates around the vertically-placed shaft.

Incorporated within the device is a small tachogenerator, the voltage output of which is proportional to the rotational speed.

APPLICATIONS This type of device is generally for non-meteorological use but where wind speed has to be taken into account.

Courtesy Brookes & Gatehouse Ltd

RANGE LIMITS 0.3 m/s to 50 m/s in a single instrument is typical.	
ERROR ±4% of full range or better is readily available.[1]	
OVER-RANGE PROTECTION Well in excess of 50 m/s. They will survive most wind conditions, subject to the reliability of the supporting structure.	
SUPPLY VOLTAGE Self-generating	
SENSITIVITY 0.15 V dc per m/s as an example.	
CROSS SENSITIVITY Cup anemometers respond to air flows perpendicular to the plane of rotation. Do not use them where a vertical component of flow is present.	
TEMPERATURE RANGE Operating: −10°C to +60°C or better.	
TEMPERATURE SENSITIVITY Insignificant.	
SIGNAL CONDITIONING Depends upon the use to which the device is being put but it is usually advisable to use the manufacturer's system.	
BENEFITS AND LIMITATIONS The output will be sensitive to the electrical output load.	
COMMENTS [1]Although essentially linear, there is a small bearing friction which requires an initial wind flow to start rotation. However, for their normal use, low speed measurement (2 to 3 m/s) is not important. This type of device is commonly fitted with an electrical warning device incorporated within the manufacturers signal conditioning to identify when the wind speed reaches some predetermined level(s).	
SUPPLIERS BRO; TEL.	

FG5. Unrestrained flow, hot wire/film anemometer

OPERATING PRINCIPLES A metal wire or film, usually tungsten, platinum or platinum-iridium, is heated and then subjected to a cooling flow. The rate of heat transfer from wire to fluid will be related to the fluid velocity.

The wire is heated from a constant voltage source and the variation in current is measured. More sophisticated systems use a bridge circuit which automatically maintains the wire at a constant temperature and measures the bridge output voltage.

APPLICATIONS Used mainly for research and development projects. Well suited for high temperature flows. For first users it is wise to discuss any application with the manufacturers prior to purchase.

Courtesy BRITAL

RANGE LIMITS Gases, hot wire/film: 15 to 350 m/s.[1] Liquids, hot wire/film: 0.03 to 30 m/s.[1] Frequency response Gases: hot wire 600 kHz, hot film: 300 kHz.[2]

ERROR Difficult to define.[3]

OVER-RANGE PROTECTION The velocity values quoted are maxima, especially if the fluid contains solid particles. They must be handled with great care.

SUPPLY VOLTAGE Operating resistance: 4 to 15 Ω. Operating currents: up to 0.35 A (air), 0.8 A (liquids) for typical probe requirements.

SENSITIVITY Tungsten: 0.0042 Ω/°C.[4] Platinum: 0.003 Ω/°C.[4] Platinum-iridium: 0.0009 Ω/°C.[4] (The lower the value, the higher the maximum temperature.)

CROSS SENSITIVITY These transducers have a degree of transverse sensitivity which will be determined by calibration.

TEMPERATURE RANGE Up to +300°C for film and +800°C for wire.

TEMPERATURE SENSITIVITY Compensation is required for the local fluid temperature.

SIGNAL CONDITIONING A great degree of sophistication is possible. It is recommended that prior to commencing a measurement task, the manufacturer is consulted.

BENEFITS AND LIMITATIONS Available as single or multidirectional probes. Their use is specialised and require both skill and experience to achieve good quality results.

COMMENTS A device of fast response capable of measuring turbulence intensity and Reynolds stress. [1]Typical values. These devices come in a wide array of choices and will operate up to supersonic speeds. [2]Dictated by the quality of the signal conditioning and the accuracy to which the velocity calibration can be achieved. [3]The performance in liquids is generally better than for gases but thermal boundary layer effects may degrade the actual performance. [4] These devices are nonlinear.

SUPPLIERS BIR; TES; TSI1; TSI2; TSI3.

Flow, gases

FG6. Unrestrained flow, rotating vane anemometer, capacitive

OPERATING PRINCIPLES These devices consist of a multi-bladed metal axial impeller protected by a duct.
The passage of each blade past a location creates a change in capacitance of a metal plate located in the fan duct.
The resultant pulses are counted and are a measure of the velocity.

APPLICATIONS Often hand-held they are used for flow in rooms, ducts, pipework and ventilation systems. They are also used for studies of localised wind flow around buildings and large structures and in wind tunnels.

Courtesy Airflow Developments Ltd

RANGE LIMITS Flow: 0.2 to 30 m/s.

ERROR Accuracy: ±1% of reading or better with dedicated calibration.

OVER-RANGE PROTECTION Capable of accepting some over ranging but an operator could be at considerable risk at the higher flows.

SUPPLY VOLTAGE Portable: 6 V dc battery. Permanent installations: mains.

SENSITIVITY Three seconds is a minimum period to give a reliable averaging.[1]

CROSS SENSITIVITY Not sensitive to cross flow. Care must be taken to ensure that the operator or other extraneous objects do not disturb the local flow.

TEMPERATURE RANGE –30°C to +70°C.

TEMPERATURE SENSITIVITY Insignificant.

SIGNAL CONDITIONING Self-contained and linked to the measuring head by an electrical cable. Display can be either analogue or digital.

BENEFITS AND LIMITATIONS These devices are easy to use but are direction-sensitive thus care must be exercised to ensure a correct reading.

COMMENTS A most traditional way of measuring air flow.
[1]Up to two minutes will provide better accuracy particularly for lower velocities.

SUPPLIERS AIR.

Flow, gases

FG7. Unrestrained flow, rotating vane anemometer, inductive

OPERATING PRINCIPLES These devices consist of a multi-bladed metal axial impeller protected by a duct.
The passage of each blade past a location creates a change in inductance of a coil located in the duct housing.
The resultant electrical pulses are counted and are a measure of the velocity.
Devices may contain additional measurements such as temperature or humidity (RH).

APPLICATIONS Often hand-held, they are used for flow in rooms, ducts, pipework and ventilation systems. They are also used for studies of localised wind flow around buildings and other large structures.

Courtesy Testoterm Ltd

RANGE LIMITS Velocity: 0.4 to 40 m/s.
ERROR Accuracy: ±1% of reading or better with dedicated calibration.
OVER-RANGE PROTECTION Capable of accepting some over ranging but an operator could be at considerable risk at higher flows.
SUPPLY VOLTAGE Portable: 6 V dc battery. Permanent installations: mains.
SENSITIVITY Analogue output: 20 mV/m/s.[1]
CROSS SENSITIVITY Not sensitive to cross flow. Care must be taken to ensure that the operator or other extraneous objects do not disturb the local flow.
TEMPERATURE RANGE Operating: 0 to +40°C. Storage: –30°C to +80°C.
TEMPERATURE SENSITIVITY Insignificant.
SIGNAL CONDITIONING Self-contained and linked to the measuring head by an electrical cable. Displays are analogue, digital and by printer.
BENEFITS AND LIMITATIONS These devices are easy to use but are direction-sensitive thus care must be exercised to ensure a correct reading.
COMMENTS The choice between this and the capacitive device, described at FG6, is unlikely to be affected by the type of sensing. They are comparable and competitive. [1]Three seconds is a minimum period to give a reliable averaging with up to two minutes for better accuracy, particularly for lower velocities.
SUPPLIERS EUR1; OCE; TES.

Flow, gases

FG8. Unrestrained flow, rotor, vertical wind speed

OPERATING PRINCIPLES These devices are converted from normal cup anemometers with the cups replaced by a multibladed axial rotor.
This rotor responds to the vertical component of wind speed whilst being immune to the horizontal component.

APPLICATIONS Generally mast-mounted, this device is a useful adjunct when studying the behaviour of wind disturbance around buildings and structures.

Courtesy R.M. Young & Co.

RANGE LIMITS	0.2 m/s to 50 m/s.
ERROR	±4% of full range.[1]
OVER-RANGE PROTECTION	Vertical wind velocities in excess of 50 m/s are unlikely.
SUPPLY VOLTAGE	Depends upon anemometer. (See FG3 and FG4.)
SENSITIVITY	Again dependent upon the choice of anemometer.
CROSS SENSITIVITY	Low. Axial rotors lack response to transverse flow. Their inertia limits speed of response to fluctuations of flow (turbulence).
TEMPERATURE RANGE	Limited only by the anemometer.
TEMPERATURE SENSITIVITY	Insignificant.
SIGNAL CONDITIONING	As required by the main anemometer device.[2]
BENEFITS AND LIMITATIONS	A simple way in which to measure the vertical component of wind velocity subject to the average value being sufficient.
COMMENTS	A convenient conversion but calibration may be difficult since the device may not perform properly with its rotational axis horizontal – a common situation if wind tunnel calibration is used.

[1]This is dependent upon the original cup anemometer used for the conversion.
[2]Since it is likely that the use of this instrument will also require conventional wind speed measurement, there is merit in using a common signal conditioning system for both types of measurement.

SUPPLIERS	RMY.

Flow, gases

FG9. Unrestrained flow, thermal anemometer, thermistor

OPERATING PRINCIPLES A thermistor, located in the air flow, has its temperature maintained at some elevated value. Air passing over the thermistor will provide a cooling effect and the required change of electrical current to maintain the temperature is related to the flow velocity. To correct for the local ambient temperature, a correction is applied by measuring the ambient temperature with a second thermistor whose output is used to provide a correction signal.[1]

APPLICATIONS Usually hand-held, flow and temperature may be measured from the one instrument.
Well suited for heating and ventilating engineering, the small probe diameter allowing easy access through ventilator grills.

Courtesy Airflow Developments Ltd

RANGE LIMITS Flow: 0 – 2 m/s to 0 – 50 m/s. Temperature: 0°C to +80°C.

OVER-RANGE PROTECTION Capable of accepting extremely high over-speeding. For portable devices, the operator could be at considerable risk at higher flows.

ERROR Flow accuracy: ±2% of full range.[2,3] Temperature: ±3% of full range.

SUPPLY VOLTAGE Being portable these devices are battery operated.

SENSITIVITY Dependent upon the signal conditioning system. See also TE5.

CROSS SENSITIVITY Insignificant, since these devices are not affected by flow direction.

TEMPERATURE RANGE Compensated: –10°C to +90°C.

TEMPERATURE SENSITIVITY Temperature compensated. High humidity may introduce some error.

SIGNAL CONDITIONING Usually self-contained with the thermistor probe either connected by cable or directly mounted to the display and signal conditioning unit.[3]

BENEFITS AND LIMITATIONS An interesting technique which provides a device which is essentially free of directional error and allows measurement at a "point" rather than an area.

COMMENTS Some probe arrangements may be telescopic, up to 1 metre in length, giving good remote access. [1]Hence some systems may be used to measure both flow and temperature. [2]Better accuracy is possible with dedicated calibration. [3]The relationship between the heat transfer and flow is a nonlinear one. Instruments may contain linearising circuitry and both analogue and digital displays are available. Being nonlinear they have a better resolution at the lower velocities making them particularly suitable for low flow measurement.

SUPPLIERS AIR; BIR; EUR1; TES.

Flow, gases

FG10. Unrestrained flow, ultrasonic anemometer

OPERATING PRINCIPLES An ultrasonic signal is sent between two piezoelectric crystals.
Wind flow along this same path will hasten or slow down the ultrasonic transmission velocity, thus the time of flight of the signal is a measure of the wind speed. By using three axes the true wind vector can be established.[1] Because timed measurements are made in both directions and the time difference used to establish the velocity, the system is independent of the speed of sound in the gaseous medium.

APPLICATIONS Usually mast-mounted, they are for specialised applications aimed at providing a detailed analysis of the wind structure or the effects of local obstructions.

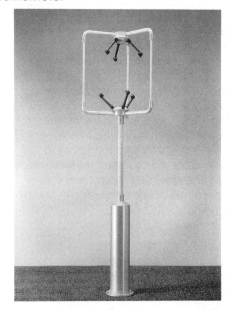

Courtesy BRITAL

RANGE LIMITS Speed: 0 to 60 m/s.[2] Direction: 0 to 360°, all axes.

OVER-RANGE PROTECTION None necessary for meteorological uses except in ensuring that the supporting structure is secure and not at risk from flying debris.

ERROR Horizontal wind speed accuracy: ±1.5% of full range, ±0.02 m/s and ±0.1% of full range/° of elevation. Vertical wind speed ±3% of full range, ±0.02 m/s. Wind direction ±2° (1 to 30 m/s) or ±5°. (0.5 to 60 m/s).

SUPPLY VOLTAGE 19 to 30 V dc or mains ac supply.

SENSITIVITY Up to 5 V dc for full range.

CROSS SENSITIVITY There will be some transverse sensitivity and thus may require calibration.[3]

TEMPERATURE RANGE Operating: –20°C to +50°C.[4] Storage: –40°C to +75°C

TEMPERATURE SENSITIVITY Essentially free of temperature effects.

SIGNAL CONDITIONING Purpose-built equipment specific to the transducer.

BENEFITS AND LIMITATIONS Measures turbulent flow. (Typical response frequency is 1 Hz). Because of its operation, direct comparison of measurement with cup anemometers is not possible.

COMMENTS [1]Since the transmission time is short, a measure of turbulence can be made although the "size" is limited to the spacing between associated crystals. [2]They can be supplied with an option of temperature measurement, again using acoustic techniques. A typical temperature range is –10°C to +50°C. [3]As much of the final support structure as possible should be included during calibration. [4]It is essential to keep the crystal elements free from ice.

SUPPLIERS BIR; GIL.

Flow, gases

FG11. Unrestrained flow, wind direction, potentiometer

OPERATING PRINCIPLES A wind direction vane comprises a horizontal rod or beam, with a vane located at one end, and attached at its balance point to a vertical shaft.

This vertically-mounted shaft, running in low friction bearings, is connected to the wiper of a resistance potentiometer.

When subject to a horizontal wind the vane will cause the beam to face into the general wind direction and this position will be identified by the potentiometer reading.

APPLICATIONS Mast-mounted. Occasionally used alone but more commonly in tandem with a cup anemometer; providing a total system to determine wind velocity.

Courtesy Brookes & Gatehouse Ltd

RANGE LIMITS 0 to 360° continuous.[1]	
ERROR Nonlinearity: ±0.5% of full range. Resolution: ±0.2% of full range. Repeatability: ±0.5°.	
OVER-RANGE PROTECTION These devices are constructed to operate in the same conditions as the cup anemometers – in excess of 75 m/s.	
SUPPLY VOLTAGE Typically 20 V dc (max) to the potentiometer.	
SENSITIVITY For the quoted supply voltage above: 0.06 V/°.[2]	
CROSS SENSITIVITY Insignificant if properly mounted and balanced.	
TEMPERATURE RANGE Operating: –50°C to +70°C.	
TEMPERATURE SENSITIVITY 0.005% of full range/°C or better.	
SIGNAL CONDITIONING A simple voltage measurement is required but it is preferable to use equipment provided by the supplier.	
BENEFITS AND LIMITATIONS Perhaps the simplest of all the types of the available wind vanes.	
COMMENTS [1]But with up to 2° uncertainty at the 0/360° overlap position. (Usually identified as "North"). [2]These devices require a threshold level of wind speed to initiate angular movement; 1 m/s is an example.	
SUPPLIERS BRO; OCE; TEL; VEC.	

Flow, gases

FG12. Unrestrained flow, wind direction, reed switching

OPERATING PRINCIPLES The wind vane comprises a horizontal rod or beam, with a vane located at one end, and attached at its balance point to a vertical shaft running in low friction bearings.

Its angular position is identified in sectors by equispaced reed switches and a magnet.[1]

These switches form either part of a resistive potential divider network or they are used to control an electronic encoder circuit.

APPLICATIONS Mast-mounted and used in conjunction with a cup anemometer. This type of device is most commonly used for meteorological measurements.

Courtesy Vector Intruments

RANGE LIMITS 0 to 360° continuous, with an overlap of up to 360°.[2, 3]

ERROR Instantaneous: sector angle. Continuous: ±2° or better with filter.

OVER-RANGE PROTECTION These devices are constructed to operate in the same conditions as the cup anemometers – in excess of 75 m/s.

SUPPLY VOLTAGE These systems come complete with their own signal conditioning and are generally low voltage dc.

SENSITIVITY Analogue output: ±6.5 V dc. (0.036 V/°). Encoded output: 4-bit Gray code.

CROSS SENSITIVITY Insignificant if properly mounted and balanced.

TEMPERATURE RANGE Operating: –55°C to +70°C.

TEMPERATURE SENSITIVITY 0.025% of full range/°C.

SIGNAL CONDITIONING Either voltage or encoded output using the manufacturer's system complete with a visual output, analogue or digital, and an output socket.

BENEFITS AND LIMITATIONS These devices have a good reputation and are used for many meteorological applications.

COMMENTS [1]Although the output operates in relatively coarse steps, the turbulent wind flow causes the vane to be in continuous movement across the sectors. By providing a continuous averaging of the output, the resolution is improved considerably.
[2]To provide transition over the change-over point, 0/360°, the potentiometric network circuitry has considerable overlap.
[3]These devices require a threshold level of wind speed to initiate angular movement; typically 0.3 m/s.

SUPPLIERS VEC.

Flow, liquids

The early engineers, using the elegant relationship defined by Bernoulli, made more simple by being concerned with incompressible flow, provided methods of measurement which were both visible and measurable by simple passive means. Flow, for example, over the "V notch" or an open Venturi enabled measurement by height or liquid level.

More modern demands changed this simple approach and now considerable effort has been made to develop devices which record rapidly, efficiently and often without interference to the flow.

For flow in pipes, the devices are mostly pre-assembled into pipe sections of the preferred size and equipment comes complete with its own signal conditioning and added facilities such as alarm warnings.

It is important that flow measurement is made away from a location where there is a risk that the flow will experience some adverse influence such as bends, contractions or discontinuities. As a working rule the flow should experience a stable uniform flow for a length of at least five and preferably ten pipe diameters away from any bends, area changes or other flow restriction upstream of the device, and perhaps half these values downstream. For some devices, where there is risk of air entrainment rising to the top of the pipe, it is preferable to make the measurement in a vertically mounted mode.

Even with the introduction of electrical devices, some of those presented overleaf remain essentially of the style of the older mechanical designs, updated only in that they are an adaptation. Indeed there are a significant range of devices available which are entirely mechanical including the drive to the read-out display. However where remoteness or production control is essential, the electrical device comes into its own.

Pipe flow is usually measured in either volume flow (litres per minute) or by velocity (metres per second) depending on the type of sensing. Direct measurement in pipe flow requires that the sensing integrates at least across the pipe diameter.

The most accurate calibration method for pipe flow devices is to provide a test "circuit" by which the liquid may be set in motion under controlled conditions and then suddenly diverted for a known period. This diverted water is caught in a container and then usually weighed to measure its mass and then calculate the volume.

For unrestrained flow, such as is used for flow in rivers or at sea, the preferred methods remain the rotor type current meters although the demands for the study of flow in waves has given rise to electromagnetic techniques.

Calibration of devices used for unrestrained flow is achieved by moving the device through still water at a known speed. These calibration tanks are not common and very few companies have dedicated facilities. It is important that these tests are performed with towing equipment identical to that which is to be used later. Current meters can be extremely sensitive to local flow variations set up by the towing gear.

Wherever possible it is advisable to purchase equipment complete with its calibration information. The only time that calibration should be attempted is where greater accuracy is an essential requirement.

NAMAS Accredited Laboratories: 0009, 0031, 0060, 0097, 0134 and 0255.

Contents: Flow, liquids

FL1. Pipe flow, electromagnetic
0.5 to 11 m/s.

FL2. Pipe flow, helical screw
30 to 200 l/min.

FL3. Pipe flow, rotating vane (Pelton wheel type)
Direct flow: 0.25 to 220 l/min.

Bypass flow: 350 to 20 000 l/min.

FL4. Pipe flow, ultrasonic (Doppler)
Up to 10 m/s.

FL5. Pipe flow, ultrasonic (time of flight)
Up to 15 m/s.

FL6. Unrestrained flow, current meter, axial rotor
Up to 15 m/s.

FL7. Unrestrained flow, current meter, electromagnetic
Up to 18 m/s.

FL8. Unrestrained flow, current meter, Savonius rotor
Up to 3 m/s.

FL9. Unrestrained flow, ship's speed, electromagnetic
−4 to +30 kn (knots).

FL10. Unrestrained flow, ship's speed, paddle wheel
Up to 30 kn.

NOTE. Some devices are suitable for both gas and liquid flow. Where this is so, they are described under Flow, gases, and these are:

FG1. Pipe flow, mass flow, temperature.

FG2. Pipe flow, variable area.

FG5. Unrestrained flow, hot wire/film anemometer.

FL1. Pipe flow, electromagnetic

OPERATING PRINCIPLES These transducers exploit Faraday's Law which states that when an electrical conductor (the liquid) passes through a magnetic field the voltage induced within the conductor is proportional to its velocity.

The device is essentially a transformer with the two windings separated across the pipe diameter and two electrodes which connect to the induced voltage. The liquid must be a conductor. For these purposes aqueous solutions are suitable whereas hydrocarbons are not.

APPLICATIONS Supplied as a complete pipe and system, they are widely used for water flow especially where pulsating or erratic flow conditions prevail.

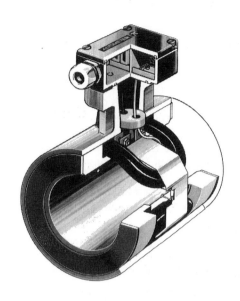

Courtesy Krohne Measurement & Controls

RANGE LIMITS Range: 0 – 0.5 m/s to 0 – 11 m/s.[1]

ERROR Accuracy: ±0.2% of full range (calibrated); ±0.5% of full range or ±1.0% of reading (uncalibrated).

OVER-RANGE PROTECTION Will accept considerable excess flow at pressures ranging from 0 Pa (absolute) to 25 MPa (subject to lining material).

SUPPLY VOLTAGE These devices are complete systems and hence are usually supplied for operation by mains voltage.

SENSITIVITY Voltage: typically ±5 V (forward and reverse flow). Current loop: 0 – 10, 2 – 10, 0 – 20, or 4 – 20 mA.[2]

CROSS SENSITIVITY Generally immune to solids within the liquid (slurry) unless the particles are of magnetic material. Error may occur for partially-filled pipes.[3]

TEMPERATURE RANGE Up to +190°C may be accommodated for some liquids.[4]

TEMPERATURE SENSITIVITY Insignificant.

SIGNAL CONDITIONING Self-contained. Energisation of the transducer is either sinusoidal, or pulsed dc. There are benefits with both options.

BENEFITS AND LIMITATIONS The measurement has no effect upon the flow and provides a true integration of the local velocities.

COMMENTS Suitable for pipe diameters 2.5 mm dia to 1.6 m in a wide range of sizes.
[1] 2 – 3 m/s is a preferred velocity or up to 5 m/s to prevent deposition of slurry.
[2] A frequency pulse is available. [3] The technique is insensitive to viscosity, pressure, temperature or density. [4] Since the preferred liquids are aqueous, 0°C is the practical lower limit; however the transducer will operate down to –25°C.

SUPPLIERS ABB2; BRI; KRO.

FL2. Pipe flow, helical screw

OPERATING PRINCIPLES Two helical screws (the rotors) are placed with their axes parallel to each other and in physical engagement.

The two screws, one of conventional thread form and the other of an Acme form, maintain intimate contact and are housed in overlapping tunnels which form the meter's body.

This provides two pockets made up from the rotors and body walls into which the liquid flows causing rotor rotation which is measured by an inductive pick-up. (See VA2).

APPLICATIONS Fitted into a pipework system, its design is well-suited to the monitoring of high viscosity fluids such as oils, acids and food processing liquids.

Courtesy Litre Meter

RANGE LIMITS 0.5 – 30 l/min to 25 – 2000 l/min.[1]	
ERROR Accuracy: ±0.1% of measured value. Repeatability: ±0.01% of measured value.	
OVER-RANGE PROTECTION ×1.5 of full range.[2]	
SUPPLY VOLTAGE A typical range of low voltage dc or a choice of mains voltages.	
SENSITIVITY Voltage: 10 V for full range. Current: 0 – 20 or 4 – 20 mA current loop.[3]	
CROSS SENSITIVITY Insignificant but relies on lubrication by the liquid flow.	
TEMPERATURE RANGE –20°C to +230°C.	
TEMPERATURE SENSITIVITY Insignificant.	
SIGNAL CONDITIONING Supplied as a complete system with a visual display, analogue signal alarms and a totaliser count.	
BENEFITS AND LIMITATIONS A positive displacement device free of most of the effects of viscosity but there will be some pressure drop across it.	

COMMENTS Capable of accepting both high and variable viscosity fluids with a precision performance.
[1]The pressure rating is dependent upon the body material but ranges from 4 MPa to 35 MPa. Pipe sizes range from 19 mm to 100 mm.
[2]It is likely that excessive pressure drop across the device will make the flow somewhat self-limiting.
[3]The rotor-generated frequency pulse is also available and ultimately this controls the sensitivity.

SUPPLIERS LIT; PLA.

Flow, liquids

FL3. Pipe flow, rotating van (Pelton wheel type)

OPERATING PRINCIPLES A multibladed Pelton wheel is mounted on an axle which in turn is mounted in low friction bearings.
The liquid being monitored passes through an inlet jet, impinges upon the blades of the wheel, causing rotation, and then exits through the outlet.
Each blade tip carries a magnet (ferrite) which pass an externally-mounted inductive pick-up.
The resultant pulses are used to measure the flow rate.[1]

APPLICATIONS Connected by normal pipe fittings, they are capable of operating in a wide range of liquids including corrosive ones such as sulphuric or hydrochloric acids.

Courtesy Litre Meter

RANGE LIMITS Direct flow: 0.01 – 0.25 l/min to 5 – 220 l/min.[2] Bypass flow: 7 – 350 l/min to 500 – 20 000 l/min.

ERROR Nonlinearity: ±0.5% of full range. Repeatability: ±0.25% of full range.

OVER-RANGE PROTECTION The values identified are considered maximum ones although one would expect them to accommodate some over speeding for a short period.

SUPPLY VOLTAGE A typical range of low voltage dc or a choice of mains voltages.

SENSITIVITY With signal conditioning. Voltage: 10 V for full range.[3]
Current: 0 – 20 mA or 4 – 20 mA.

CROSS SENSITIVITY Insignificant, but see Note 1, below.

TEMPERATURE RANGE Up to + 135°C.

TEMPERATURE SENSITIVITY Insignificant, except for possible effects from changes in liquid viscosity.

SIGNAL CONDITIONING Supplied as part of a system complete with a visual display, alarms, a signal output and a totaliser count.

BENEFITS AND LIMITATIONS Viscosity-sensitive. For low viscosity the effects are minimal but for higher values the rotor will experience a fluid drag error.

COMMENTS [1]For larger flows the device is installed into a bypass whereby only part of the direct flow is monitored. In such cases care must be exercised to ensure a representative flow through the bypass.
[2]The pressure rating is dependent upon the body construction with typical values of pressure ranging between 0.5 MPa to 30 MPa in pipe sizes between 6.4 mm and 305 mm diameter.
[3]The blade passage frequency pulse is also available.

SUPPLIERS LIT.

FL4. Pipe flow, ultrasonic (Doppler)

OPERATING PRINCIPLES Two ultrasonic crystals are mounted on the exterior surface of a metallic pipe, one upstream of the other and close to each other.

Ultrasonic signals are transmitted from one crystal, reflected by impurities within the liquid and received by the other, the frequency of the transmission being modified or shifted by the motion of the impurities.

Hence the output frequency change is a measure of the liquid flow.

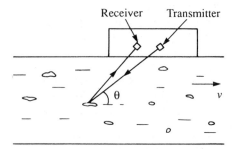

APPLICATIONS Installed by simple clamping or hand-held on to existing pipework, they provide a simple flow measuring system for a wide range of applications

Doppler flowmeter principle

Courtesy Butterworth-Heinemann Ltd

RANGE LIMITS 0.1 to 10 m/s.[1]

ERROR Accuracy: ±5% of full range or better. Repeatability: ±1% of full range.

OVER-RANGE PROTECTION The device is not in the flow path hence no protection is required.

SUPPLY VOLTAGE Portable, battery-operated devices.

SENSITIVITY The sensitivity will depend upon the number of display digits.

CROSS SENSITIVITY The system is dependent upon impurities being present and these can include suspended solids (slurry), bubbles or turbulence in the flow path.

TEMPERATURE RANGE +5°C to +115°C.

TEMPERATURE SENSITIVITY Insignificant.

SIGNAL CONDITIONING Battery-operated self-contained device, complete with a visual display panel.

BENEFITS AND LIMITATIONS It is portable, non-intrusive and very easy to install but does require impurities to be present in reasonable quantity.

COMMENTS Uses a true velocity measuring technique. These devices are portable installations allowing the rapid *ad hoc* measurement of flow wherever and whenever required. It is important that the pipe is full even under "no flow" conditions.
[1]Recommended minimum pipe size is 20 mm diameter.

SUPPLIERS EUR1; UCC1; UCC2; UCC3.

Flow, liquids

FL5. Pipe flow, ultrasonic (time of flight)

OPERATING PRINCIPLES Two crystals are mounted on the exterior surface of a metallic pipe such that one is upstream of the other and separated by 180° rotation.

Ultrasonic signals are transmitted alternatively from each probe and received by the other.

The time of flight of each signal will include the effects of the liquid motion and by combining the two measurements, the velocity of the liquid can be established.[1]

APPLICATIONS Attached by banding clamps, these provide both a temporary or permanent installation for the measurement of pipe flow for any liquid without interfering with the flow.

Courtesy KDG Mowbrey Ltd

RANGE LIMITS 0.3 m/s to 15 m/s, influenced by the pipe diameter.[2]

ERROR Accuracy: ±1% of flow value (permanent installation).[3] Nonlinearity: ±1% of flow value. Repeatability: ±0.5% of flow value (in excess of 0.3 m/s).

OVER-RANGE PROTECTION The system cannot be damaged by flow in excess of the maximum designed values.[4]

SUPPLY VOLTAGE Being self-contained, the system will utilise a local low voltage dc supply or mains.

SENSITIVITY After signal conditioning, typical outputs are a digital display, 4 – 20 mA current loop and a signal frequency of 0 – 1 kHz.

CROSS SENSITIVITY The liquid should be free of air bubbles or solid particles (less than 1% by volume) and totally filling the pipe.

TEMPERATURE RANGE –50°C to + 130°C

TEMPERATURE SENSITIVITY Signal conditioning: ±0.025% of full range/°C. Liquid, zero: 0.01% of full range/°C. Liquid, range: 0.05% of full range/°C.

SIGNAL CONDITIONING Supplied complete as part of the whole system.

BENEFITS AND LIMITATIONS This system can be fitted retrospectively to existing metal pipes for both permanent and temporary (kit) installations.

COMMENTS [1]The "speed of sound" term is eliminated from the equation but allowance has to be made for the angle of the signal path to the flow path.
[2]Recommended for pipe sizes of between 25 mm and 2000 mm.
[3]For kit systems, the accuracy may degrade to between ±2.5 and ±5.0% of flow value. (The repeatability and nonlinearity are unchanged).
[4]Damage may occur to the crystal elements if no liquid is present.

SUPPLIERS KDG; KRO.

Flow, liquids

FL6. Unrestrained flow, current meter, axial rotor

OPERATING PRINCIPLES A multibladed rotor is mounted upon an axle and projected from the front of a body which incorporates a directional vane at its rear.

When subject to a water flow, the body points the rotor into the flow which initiates the rotation of the rotor.

This is measured by an arrangement of magnets and a reed switch. (Some devices may use the Hall effect to provide the revolutions count).

APPLICATIONS Either supported by wading rods (shallow) or by cables (deep) they are most commonly used for river and ocean water flow under steady conditions.

Courtesy Valeport Marine Scientific Ltd

RANGE LIMITS 0.025 m/s to 10 m/s.[1]

ERROR Accuracy: ±1.5% of full range.[2]

OVER-RANGE PROTECTION In normal use it is unlikely that the upper range value will be encountered.

SUPPLY VOLTAGE Sufficient to receive a suitable output from the reed switch.

SENSITIVITY Typically one or two pulses per revolution.

CROSS SENSITIVITY Insignificant in steady flow but in wave conditions the results may be of little or no value.[3]

TEMPERATURE RANGE −5°C to +45°C.

TEMPERATURE SENSITIVITY Essentially free from temperature effects.

SIGNAL CONDITIONING A counter displays either the number of revolutions over a preset (adjustable) time or records the time for a preselected number of revolutions.[4]

BENEFITS AND LIMITATIONS These devices when hung from a suspension cable require a suitable swivel to allow 360° rotation. This may give rise to problems of signal transmission to the water surface. A sinker weight will be needed to maintain the required depth in high flow conditions.

COMMENTS [1]Bearing friction dictates the velocity threshold. An upper range value of 15 m/s can be provided for special requirements.
[2]These devices are not linear because of bearing friction. Normally a calibration is provided either as a series of equations or a look-up table. Better results may be obtained by individual calibration.
[3]Inaccurate measurement can arise due to fouling of the rotor by sea weed.
[4]Dictated by the flow velocity and final purpose of the measurement.

SUPPLIERS VAL.

Flow, liquids

FL7. Unrestrained flow, current meter, electromagnetic

OPERATING PRINCIPLES This exploits Faraday's Law which states that when an electrical conductor passes through a magnetic field the voltage induced within the conductor is proportional to its velocity.
The device is essentially a transformer with the windings being so positioned that the transformer coupling is modified in proportion to the flow around the windings.[1]

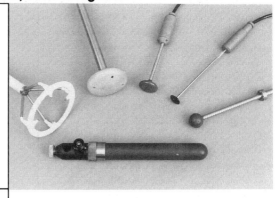

APPLICATIONS Rigidly mounted from a supporting structure, they are used to measure velocities in both steady and turbulent water flow.
The transducers are disc-shaped, annular or spherical in shape.

Courtesy Valeport Marine Scientific Ltd

RANGE LIMITS 0 to 18 m/s.

OVER-RANGE PROTECTION None required.[2]

ERROR Disc: nonlinearity ±5 mm/s of measured value; heading error ±5° of measured value; Annular: nonlinearity ±2 mm/s of measured value; heading error ±3° of measured value. Spherical: nonlinearity ±20 mm/s of measured value; heading error: ±5° of measured value.

SUPPLY VOLTAGE Complete with its own signal conditioning. Both low voltage (dc) supplies and mains (ac) supply versions are available.

SENSITIVITY 5 mm/s (disc); 10 mm/s (annular); 20 mm/s (spherical).[3]

CROSS SENSITIVITY Transverse sensitivity determined by calibration..

TEMPERATURE RANGE Typical river/sea temperatures.

TEMPERATURE SENSITIVITY Insignificant.

SIGNAL CONDITIONING Supplied as part of the full system including the energising ac voltage, a visual display unit and output signal.

BENEFITS AND LIMITATIONS Each shape has a best use. Disc for two-axis steady flow, annular for two-axis unsteady flow and spherical for three-axis flow.

COMMENTS [1]Water motion is often multiaxis. To measure this flow faithfully and without local modification, the flow meter shape is of great importance. Originally disc-shaped, these stall in one axis at small flow angles. Later these sensors were extended to both annular and spherical shapes.
[2]Flow patterns which are smaller than the physical dimension of the sensor head may give ambiguous results. A filter is incorporated to eliminate such higher frequency components. [3]Dictated by the electrical noise level.

SUPPLIERS OCE; VAL.

Flow, liquids

FL8. Unrestrained flow, current meter, Savonius rotor

OPERATING PRINCIPLES Comprises two equispaced curved vanes or blades fixed to a vertically-mounted axle shaft.
Water flowing across the curved surfaces causes the rotor to spin.
(Simple Savonius rotors are seen commonly as advertising devices outside of shops and garages, attracting the viewer by their rotation).
By measuring the speed of rotation, using typically reed switches and magnets, a direct measure of flow is obtained.
Unlike axial rotors they are not vector-sensitive.

APPLICATIONS Cable-suspended or rigidly mounted, they are used for measuring sea water flow; particularly at depths as great as 6000 m.

Courtesy Oceano Instruments (UK) Ltd

RANGE LIMITS 0.02 m/s to 3 m/s.[1]

OVER-RANGE PROTECTION ×2 of full range, subject to the risk to the auxiliary suspension or mooring equipment.

ERROR Nonlinearity: between ±0.5% and ±1.0 % of full range.[2]

SUPPLY VOLTAGE The reed switches require a small dc voltage supply.

SENSITIVITY Typically one or two pulses per revolution.

CROSS SENSITIVITY Insignificant, subject to the rotor axis being vertical.[3]

TEMPERATURE RANGE −5°C to +35°C.

TEMPERATURE SENSITIVITY Essentially free from temperature effects.

SIGNAL CONDITIONING Comes complete with either a linked or inbuilt data logging system.[4]

BENEFITS AND LIMITATIONS Generally insensitive to flow direction. Where unsteady flow conditions are present it has a better performance than other rotor devices.

COMMENTS [1]Bearing friction will inhibit initial rotation. Hence the lower "breakout" velocity of 0.02 m/s.
[2]Subject to calibration.
[3]Inaccurate measurement can arise due to fouling of the rotor by sea weed.
[4]They may incorporate other measurements such as direction, temperature pressure and conductivity.

SUPPLIERS OCE.

Flow, liquids

FL9. Unrestrained flow, ship's speed, electromagnetic

OPERATING PRINCIPLES This exploits Faraday's Law which states that when an electrical conductor passes through a magnetic field the voltage induced within the conductor is proportional to its velocity.

In this case the conductor is the local water and the magnetic field is generated by a winding in the sensor and the induced voltage is detected by a second one, the whole unit just protruding through the skin of the ship's hull.

APPLICATIONS Specialist-mounted, this system is used for ship speed measurement.

Generally reserved for ships in excess of 1000 tonnes for both merchant and naval ships.

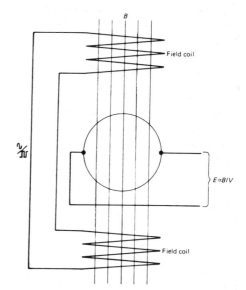

Principle of Operation. B = Magnetic field; I = Length of conductor; E = E.M.F.; V = Flow rate.

Courtesy Butterworth-Heinemann

RANGE LIMITS −3 kn to + 20 kn and −4 kn to + 30 kn (typical).[1]

OVER-RANGE PROTECTION Unlikely to experience over-speed. However they are generally immune.

ERROR Accuracy: ±0.2% of full range.[2]

SUPPLY VOLTAGE Most likely to operate on any commonly-available ac mains voltages.

SENSITIVITY 0 to 20 mA for full range.

CROSS SENSITIVITY These devices are particularly insensitive to transverse flow.[3]

TEMPERATURE RANGE Suitable for all sea temperatures.

TEMPERATURE SENSITIVITY Generally immune to temperature effects.

SIGNAL CONDITIONING Part of a complete system with speed display, mileage counter, mileage pulses for interconnection to a "Sat Nav" system and additional visual displays.

BENEFITS AND LIMITATIONS The electromagnetic log has no moving parts and is not prone to fouling.

COMMENTS See also Unrestrained flow, current meter, electromagnetic, FL7.
[1] 1 kn (knots) = 0.515 m/s.
[2] This value may be adversely affected by bad location on the hull. It is advisable to calibrate the device wherever possible over a measured distance.
[3] Indeed some incorporate a second measurement to detect transverse motion or drift of the ship caused, say, by wind effects.

SUPPLIERS TEL.

Flow, liquids

FL10. Unrestrained flow, ship's speed, paddle wheel

OPERATING PRINCIPLES To the outer skin and below the water line of the ship is fitted a small paddle wheel suitably shrouded so that it experiences rotation when the boat is moving.

This rotation is detected by incorporating a number of magnets in the paddles and sensing their passing a selected location by using a Hall effect sensor.

These pulses are converted to a suitable display or counted continuously to provide a measure of distance travelled.[1]

APPLICATIONS Fitted through the hull, they are used for ship speed measurement, generally for the smaller sizes of ship or yacht.

Courtesy NASA (1992) Ltd

RANGE LIMITS 0.5 kn (knots) to up to a maximum of 30 kn.[1,2]

ERROR Accuracy: ±2% of full range.

OVER-RANGE PROTECTION Since the boat's maximum speed will be known with some certainty, a device to match this requirement should be fitted.

SUPPLY VOLTAGE 12 V or 24 V dc (boat supply).

SENSITIVITY Smallest increment on the display unit such as 0.1 kn.

CROSS SENSITIVITY Important to ensure proper installation so that the flow reaching the paddle wheel is free of local influences from the boat.

TEMPERATURE RANGE Operating: –25°C to +70°C. Storage: –30°C to +85°C (limited periods).

TEMPERATURE SENSITIVITY Essentially free from temperature effects.

SIGNAL CONDITIONING This will be part of a complete system with a speed display and possibly a mileage counter.

BENEFITS AND LIMITATIONS Being a moving part device it is easy to check on its condition such as excessive bearing friction and other damage.

COMMENTS A simple effective device which allows small vessels to measure their speed through the water but its installation location is of vital importance to ensure proper performance. It is advisable to calibrate them after installation, such as checking the performance by running tests over a measured distance. It is subject to fouling particularly when at a mooring which may cause problems.
[1]1 kn (knots) = 0.515 m/s.
[2]There will be a small "breakout" force required to initiate rotation.

SUPPLIERS NAS; TEL.

Length and related measurements

The measurement of length is surely the oldest of all measurements. This arises because the human body is, with its appendages, a mobile length standard. The "hand", "cubit" and "yard" are such examples and in recorded history these measurements were standardised by various royal decrees. Once set, these values became remarkably consistent, transcending national boundaries, wars and natural disaster, and remained fixed for considerable periods of time.

Length standards in the form of metal bars and similar physical items were created and then the standard of measurement was maintained by comparison with these items. The original physical standards were prone to changes in length and once the capability existed for the identification of such change, new more stable length standards were sought.

For the electrical transducer industry the measurement of length using the equivalent of the electronic ruler is not common although the creation of modern laser and optical techniques has permitted electronic precision measurement over a wide range.

Within the range of electrical input/output devices the measurement of length is limited and true length transducers, as defined below, are usually so designed to satisfy a specific or unusual need. Acoustic devices form one group example. However there are a range of additional "distance" measurements to which the electrical transducer industry has responded. These may be conveniently grouped into four distinct families. These are displacement, extension, position and strain. To this may be conveniently added "proximity" which although perhaps less accurate is a form of positional measurement.

It is useful to consider here the meaning of all five of these terms. Experience shows that an incorrect interpretation of the type of measurement achieved can result in errors of deduction. This is particularly so for the difference between strain and extension. Here they have been assigned the following meanings, each of which are covered in separate sections.

Displacement (DL)

This is the measurement of the distance between a selected location and a datum which is not part of the object. An example would be the relative positions of the piston and cylinder of a hydraulic ram. Displacement is concerned with change of position rather than absolute distance.

Extension (EX)

This is the measure of any positional relationship between two points on the same object. One of the most common examples of this requirement is in establishing the elastic property of a material. Thus extension is governed by a change of length when compared to a fixed and preset length value; although this initial length may not be known with any degree of precision.

Position (PO)

This is the identification of a unique location between two bodies, one of these being the sensor. Once this location is established then some other device will establish any required measurement of length, angle or displacement.

Proximity (PO)

This is generally but not exclusively a more crude identification of position with the general purpose of identifying a state; in other words, a positional "On/Off" switch.

Many proximity devices have a commonality of operation with positional transducers but it is likely that the signal conditioning will not offer the same degree of precision.

Strain (ST)

This measurement is the mathematical relationship between some identified length on a body and its change in length when subjected to some physical influence such as load or temperature. Because strain is a ratio it is non-dimensional and is independent of the original chosen length. (This discussion is further developed later.)

Length

Length, in the more conventional sense, is the measurement of the distance between any two points or two surfaces or a combination of both. In the measurement of length, no correction is needed or applied to allow for the uncertainty of the attachment position required or datum position within the sensor.

The measurement of length as distinct from displacement is very limited in the field of electrical transducers. Indeed, it is arguable whether any of the electrical transducers measure length rather than displacement as discussed above. However the guide to their inclusion within this section is based upon the ability to establish the datum position prior even to the connection of any signal conditioning. Thus pressure devices used for liquid level measurement satisfy this criteria in that the diaphragm locates precisely the position from which the measurement is made.

Normally the reason for adopting an electrical form of measurement is related to making measurements in circumstances where the information is required for further data manipulation or convenient storage or where the length parameter is changing continuously or at least regularly.

Not only are there the obvious examples of movement of solid bodies but perhaps less well known are the range of liquid level measurements either of length (height) or position. Indeed the measurement of water depth using an **echo sounder** is one of the more rare examples of truly unambiguous electrical length measurement as defined previously and a system which is extremely common.

Transducers for length measurement are included here if switching on the datum is established immediately without manipulation of the equipment, or, by virtue of transducer construction, the datum position is self-evident. However there are devices which fulfil these requirements but are included in a different section since it is considered a more fitting place. Thus acoustic devices are included under length and displacement.

Using the same criteria, those pressure devices specifically designed for liquid level measurement are included here, rather than under pressure.

Calibration of length is readily available. Modern machines of great sophistication are available and even the humble ruler or tape measure has an order of accuracy which many other types of measurement cannot match.

For a greater degree of precision beyond the mechanical aids, optical ones are available although they are considerably more expensive.

NAMAS Accredited Laboratories.

There are a very wide range of laboratories which provide some form of length calibration. To identify them individually together with their particular speciality would serve little purpose and possibly confuse since the methods adopted to provide the calibration will depend upon individual requirements. The reader is advised to contact NAMAS for advice.

Contents: Length

LE1. Acoustic, water depth (echo sounder)
0 to 300 m.

LE2. Liquid level, pressure measurement, bonded resistance strain gauge
0 to 70 m.

LE3. Liquid level, pressure measurement, tide height
To suit water location.

LE4. Magnetostrictive
300 mm to 3 m.

Length

LE1. Acoustic, water depth (echo sounder)

OPERATING PRINCIPLE A piezoelectric transducer, located beneath a water surface, such as the bottom of a ship, faces towards the seabed. On energisation it transmits a string of acoustic pulses and then "listens" for the return echo. By measuring the time interval between the transmitted and received pulses and using a knowledge of the transmission speed of sound through the medium, the distance between the acoustic device and the solid body surface is established.

APPLICATIONS Through-hull mounted, its use is from small yachts to the largest vessels afloat. Other specialist purposes include fish detection and wreck location.

Courtesy NASA (1992) Ltd

RANGE LIMITS Typical range maxima: 4 m, 12 m, 20 m, 50 m, 120 m and 300 m.[1]

ERROR Accuracy to within a few centimetres is possible.[2]

OVER-RANGE PROTECTION Unimportant. However, care must be taken not to use these devices in air where damage to the piezo-electric crystal may occur.

SUPPLY VOLTAGE Acoustic transmission: 50 W (rms) to 200 W (rms) at 50 kHz to 200 kHz. Supply: 12 V to 40 V dc.

SENSITIVITY Example 0.1 m but dependent upon display.[3]

CROSS SENSITIVITY Errors due to ship motion, other than heave, are generally small.

TEMPERATURE RANGE Operating: –5°C to +70°C. Storage: –30°C to +70°C.

TEMPERATURE SENSITIVITY Transducer is generally immune.

SIGNAL CONDITIONING Purchased as a complete unit: crystal transducer, its drive, timing, display unit and optional audible warning alarms.

BENEFITS AND LIMITATIONS Measurement errors can arise from water density changes and false echos from sudden water temperature transition.

COMMENTS [1]Proper functioning usually starts between 0.1 m and 1 m of depth. [2]Limited by the inherent error of the device to measure time interval, errors due to water density change, motion of the vessel and roughness of the seabed surface. [3]This is dictated by the form of display. For example a paper chart will not have the same resolution as a digital readout. Typical displays are 3 or 4 digits to encompass the selected range.

SUPPLIERS NAS; TEL.

Length

LE2. Liquid level, pressure measurement, bonded resistance strain gauge

OPERATING PRINCIPLE The pressure impinges upon a diaphragm which either has strain gauges bonded directly to its surface or is coupled to a strain gauged sensing element. The diaphragm is housed within a suitable casing and is usually of stainless steel.

(Although any transducer of the correct pressure range could be used, the devices described below are specially constructed for this purpose.)

APPLICATIONS These devices are available complete with suitable attachments for tank mounting.

They are commonly used for measurement of storage tank levels.

Courtesy RDP Group

RANGE LIMITS 0 – 7 kPa to 0 – 700 kPa or 0 – 0.7 m to 70 m (water depth)	
ERROR Nonlinearity and hysteresis: ±0.25% of full range.	
OVER-RANGE PROTECTION × 1.5% of full range without damage.	
SUPPLY VOLTAGE Typically 28 V dc for inbuilt signal conditioning.	
SENSITIVITY 4 – 20 mA current output for full range.	
CROSS SENSITIVITY Requires a knowledge of the density of the liquid for precise measurement.	
TEMPERATURE RANGE Compensated: + 15°C to 70°C Operating: –20°C to + 80°C.	
TEMPERATURE SENSITIVITY Zero: 0.01% of full range/°C. Range: 0.01% of full range/°C.	
SIGNAL CONDITIONING Comes complete with its own current loop output signal conditioning.	
BENEFITS AND LIMITATIONS These devices are a convenient method of determining liquid level subject to there being no significant changes in density.	
COMMENTS These devices offer a simple method of measuring liquid level with little or no intrusion into the tank or vat. Care must be taken to provide appropriate protection from inflammable liquids.	
SUPPLIERS RDP; VAL; WES.	

Length

LE3. Liquid level, pressure measurement, tide height

OPERATING PRINCIPLE In coastal areas, near rivers, the density of the water will change as sea and fresh water mix.

Two transducers are mounted in a vertical column at a known distance apart, typically 1 m, and mounted at a chosen height above the local sea bed.

The difference in the two outputs identifies the separation distance and hence provides a continuous calibration for the pressure measurement outputs from which the tide height is determined.[1]

APPLICATIONS Installation is a specialised task, often requiring divers.

These transducers were developed specifically for coastal tide height measurement.

Courtesy Ship & Marine Data Systems

RANGE LIMITS To meet tidal requirements.

ERROR Nonlinearity and hysteresis: ±0.4% of full range.[2]

OVER-RANGE PROTECTION × 1.5 of full range is typical.

SUPPLY VOLTAGE 12 to 24 V dc and 110/240 V mains.

SENSITIVITY 1 to 3 mV/V dependent upon transducer choice.

CROSS SENSITIVITY Relies upon the water density between the two transducers being representative of the general water density.

TEMPERATURE RANGE Compensated: –5°C to +30 °C.

TEMPERATURE SENSITIVITY Dependent upon the chosen transducer. (See PR1 as an example).

SIGNAL CONDITIONING Immediate representation of height in chosen units. Also the signal conditioning can provide an electrical output for further data transmission.

BENEFITS AND LIMITATIONS More simple to install and maintain than the traditional mechanical float device and without the need for a "stilling" well.

COMMENTS These devices are new in the field of tidal measurement. However they use techniques which are proven. [1]In principle any type of pressure transducer may be used but for the example here the bonded resistance strain gauge type is identified. [2]In the limit the error performance could be twice that of a single transducer. However practice suggests that this is unlikely if transducers are selected with care.

SUPPLIERS EXP; OCE; SHI; VAL.

LE4. Magnetostrictive

OPERATING PRINCIPLE A wire is stretched and housed within a nonmagnetic tube and a magnet in a ring housing is located around the tube and free to move.

A torsional strain pulse is initiated at one end of the wire and travels along it until reaching the magnet whereupon a voltage pulse is produced.

The "time of flight" between the two pulses, mechanical and electrical, determines the magnet position.[1]

APPLICATIONS Available complete with mounting attachments, they are used as the sensors for motion control systems in robotics and machine tool applications.

Courtesy Hydraulic Research Ltd

RANGE LIMITS 0 – 300 mm to 0 – 3 m.

ERROR Nonlinearity: ±0.05% of full range. Hysteresis: 0.075 mm Repeatability: ±0.1% of full range.

OVER-RANGE PROTECTION Ample over-run is available.

SUPPLY VOLTAGE ±15 V dc.

SENSITIVITY 0 to 10 V dc or 0 to 10 V dc for full range, selectable.

CROSS SENSITIVITY Insignificant. The system is immune to transverse sensitivity.

TEMPERATURE RANGE Operating: –15°C to +85°C.

TEMPERATURE SENSITIVITY Low.[2]

SIGNAL CONDITIONING Self-contained, including analogue and velocity outputs together with zero and range adjustment to fully exploit the capability.

BENEFITS AND LIMITATIONS A robust length or displacement measuring system capable of operating in a high pressure environment. (20 MPa)

COMMENTS Although the measurement is determined by a continuous series of pulses, the transmission rate is at ultrasonic speed and thus for most practical purposes may be considered as "continuous".[1] This technique is called "magnetostrictive" and exploits the principle that the pattern of a magnetic field in a magnetic material may be distorted by applying a strain to the material. [2]The magnetostrictive technique is not temperature sensitive but the components will experience length changes due to temperature effects.

SUPPLIERS HYD; LUC1; LUC2.

Displacement, linear

The practical measurement of linear displacement can be one of essential simplicity or of great complexity. The problems arise in the requirement that simple identification of the relative position between two surfaces is generally easy whereas the true determination of position in three axes may require considerable ingenuity to achieve a satisfactory solution.

The methods of measurement are either non-contacting, contacting by spring loading or by mechanical connection. Each of these techniques has an appropriate suitability.

In selecting the most suitable transducer, consideration must be given to the nature of the measurement:

Is the requirement for a point location or will some form of averaging over a small area be acceptable?

What effect would a small rotation of the measured surface have upon the quality of measurement?

Does the load input of a spring-loaded system cause unacceptable local deformation?

Is there a risk of indenting the surface by using a ball contact?

Will dirt or liquids on the surface change the output of a non-contacting device?

There are a wide range of choices but care must be exercised to ensure a satisfactory measurement.

The measurement of displacement between two solids is not the only type of measurement. Liquid level is included in the Section on "Length" but in more simple applications could be identified as a displacement measurement. Again, force measurement could be related to some hopper weighing activities but equally it could be applied to the gradual reduction of the hopper level. However, the methods discussed below are the normal accepted techniques although alternatives should always be considered for any given application.

Some devices are of high precision and small displacement and are competitive in accuracy when compared to their mechanical equivalents. Others such as those with high range or stroke serve to provide information for control purposes. The needs of modern production engineering, particularly the use of robots, has increased the demand for the measurement of displacement by remote means.

The essential benefit of these electrical devices over the mechanical equivalents is that they lend themselves to automatic control and have a fast response allowing measurement between two moving parts, subject to the system response being adequate.

All ranges are identified as operating from "zero" position. However this is an arbitrary value and is not an actual measure of the initial or zero distance between the two measuring points.

NAMAS Accredited Laboratories: 0063, 0166, 0239 and 0244.

Contents: Displacement, linear

DL1. Contacting, bonded resistance strain gauge

±1 mm to ±15 mm.

DL2. Contacting, capacitive

5 mm to 250 mm.

DL3. Contacting, linear strain conversion, bonded resistance strain gauge

6.5 mm to 100 mm.

DL4. Contacting, linear variable differential transformer (LVDT)

±0.125 mm to ±470 mm.

DL5. Contacting, linear variable differential transformer, gauging

±0.013 mm to ±25 mm.

DL6. Contacting, resistive

10 mm to 2 m.

DL7. Noncontacting, acoustic

0.7 m to 3.5 m

DL8. Noncontacting, capacitive

0.05 mm to 26 mm.

DL9. Noncontacting, eddy current

0.5 mm to 80 mm.

DL10. Noncontacting, inductive

0.8 mm to 35 mm.

Displacement, linear

DL1. Contacting, bonded resistance strain gauge

OPERATING PRINCIPLE Strain gauges are bonded to flexible metal elements which when deformed will produce an appropriate strain detected by the strain gauges, generating an electrical output directly related to the displacement.

The flexible elements have a variety of shapes including cantilever, ring and "U" shapes.

APPLICATIONS Attachment is dependent upon use.

Best suited to low-accuracy requirements where the simplicity and versatility is beneficial.

Occasionally used as a combined force and displacement transducer.

Courtesy Techni Measure

RANGE LIMITS ±1 mm to ±15 mm for full range.[1]
ERROR Nonlinearity: ±1% of full range.
OVER-RANGE PROTECTION × 1.5 of full range is a common working value.[2]
SUPPLY VOLTAGE 10 V dc is a typical maximum.
SENSITIVITY Between 2 mV/V and 3 mV/V of supply voltage.
CROSS SENSITIVITY Transverse sensitivity generally is insignificant.[3]
TEMPERATURE RANGE Operating: 0°C to +40°C.
TEMPERATURE SENSITIVITY Low or insignificant for ambient temperatures.
SIGNAL CONDITIONING Normal strain gauge amplifier systems, using either ac or dc energisation, are commonly used.
BENEFITS AND LIMITATIONS Of low cost, these devices are so simple to make they are often purpose-built in-house by the user.
COMMENTS [1]The total range may be used as a plus/minus value or as a total displacement in a single direction, for example ±1 mm or 2 mm total. [2]It is often possible to engineer a mechanical stop just in excess of full range providing greater over-range protection. Further, by forcing the device against this stop a check upon the general calibration stability may be made. [3] The natural frequency will be influenced strongly by the chosen design shape and flexibility.
SUPPLIERS ACT; TEC.

Displacement, linear

DL2. Contacting, capacitive

OPERATING PRINCIPLES A core rod or armature slides within an outer housing.

The motion between the two causes capacitor plates to displace relative to each other so that there is a change of capacitance as displacement occurs.

Such a system incorporates a differential capacitance system giving both an increase and decrease in capacitance, for a given direction of motion about the central position, providing a convenient capacitance half bridge configuration.

APPLICATIONS The free end of the armature is either spring-loaded or threaded.

Capacitance devices have a particularly good long-term stability and will operate in high magnetic or nuclear radiation fields.

Courtesy Automatic Systems Laboratories Ltd

RANGE LIMITS 0 – 5 mm to 0 – 250 mm in various ranges.

ERROR Nonlinearity and hysteresis: ±0.01% of full range.

OVER-RANGE PROTECTION Good with some overtravel incorporated.

SUPPLY VOLTAGE Devices come supplied with signal conditioning and these are energised from a range of voltages such as ac mains power.

SENSITIVITY 10 V dc for full range after signal conditioning.

CROSS SENSITIVITY Subject to proper use, transverse sensitivity is insignificant.

TEMPERATURE RANGE Operating: +5°C to 100°C.[1] Storage: +5°C to 125°C.

TEMPERATURE SENSITIVITY Temperature coefficient values are selected to match the common applications.

SIGNAL CONDITIONING Normally supplied as part of the transducer. Alternatively a precision capacitance measuring system can be used.

BENEFITS AND LIMITATIONS Like all such "armature" devices, in the fully-retracted position the transducer length will be in excess of the travel available.

COMMENTS The capacitive displacement transducers have not, until recent times, enjoyed the popularity of inductive or resistive devices. However this situation is gradually changing with the advent of high quality devices.
[1]The lower temperature limitation is related to the possibility of ice formation on the plates. Hence with suitable protection, significantly lower operating temperatures are possible.

SUPPLIERS AUT.

DL3. Contacting, linear strain conversion, bonded resistance strain gauge

OPERATING PRINCIPLE A spindle is in contact with two cantilever beams such that as the spindle is displaced it acts upon a wedge to cause bending of the cantilevers.

These are gauged to respond to the bending and thus the electrical output is directly related to the displacement.

The whole is placed within a protective housing which also provides the spindle guidance.[1]

APPLICATIONS The free end of the "armature" is either spring loaded or threaded and they are used wherever a strain gauge system is preferred.

Courtesy Apek Sales Ltd

RANGE LIMITS	0 – 6.5 mm to 0 – 100 mm.
ERROR	Nonlinearity: ±0.1% of full range.[2]
OVER-RANGE PROTECTION	Good, with some overtravel incorporated.
SUPPLY VOLTAGE	5 to 10 V ac or dc.
SENSITIVITY	Varies between 3.6 mV/V and 7 mV/V of supply.
CROSS SENSITIVITY	Generally insignificant.
TEMPERATURE RANGE	−10°C to +60°C
TEMPERATURE SENSITIVITY	Zero: ±0.01% of full range/°C or better. Range: ±0.01 % of full range/°C or better.
SIGNAL CONDITIONING	Normal strain gauge amplifier systems, using either ac or dc energisation, are commonly used.
BENEFITS AND LIMITATIONS	A somewhat unusual technique which demonstrates the versatility of both the strain gauge and modern transducer design engineers!
COMMENTS	[1]These devices have been separately identified from the other types of bonded resistance strain gauge types since these are more in keeping with, and can offer a replacement for, the other "rod and piston" type of devices described. [2]By pre-selection this can be improved to ±0.05% of full range.
SUPPLIERS	APE.

Displacement, linear

DL4. Contacting, linear variable differential transformer (LVDT)

OPERATING PRINCIPLE A linear variable differential transformer, (LVDT), consists of a central primary winding flanked by two secondary ones. These three coils are in effect two electrical transformers and their coupling can be varied by a central core being moved through the windings. By energising the primary winding with a low voltage ac the difference in output between the two secondary ones provides a direct linear measurement of the core displacement.[1]

APPLICATIONS The free end of the armature is either spring loaded or threaded and they are used in an extremely wide range of control engineering applications.[2]

Courtesy RDP Group

RANGE LIMITS ±0.125 mm to ±470 mm.

OVER-RANGE PROTECTION High and provided with a generous travel beyond the normal displacement capability.

ERROR Nonlinearity: ±0.25% of full range is typical but ±0.1% of full range is possible.

SUPPLY VOLTAGE Transducer: 3 V rms; 400 Hz to 10 kHz.[3]
Signal conditioning: 24 V dc.

SENSITIVITY Transducer: 1.4 mV/mm/V (supply) to 250 mV/mm/V.[4]
Signal conditioning: ±2 V to ±10 V.[4]

CROSS SENSITIVITY Extremely low particularly for guided systems.

TEMPERATURE RANGE Operating: −270°C to +650°C.

TEMPERATURE SENSITIVITY Zero: ±0.01% of full range/°C.
Range: ±0.01% of full range/°C.

SIGNAL CONDITIONING Incorporated within the housing of the transducer. Since the output is high, no further amplification is likely to be required.

BENEFITS AND LIMITATIONS The LVDT has a particularly stable performance mainly determined by the performance of the magnetic coupling between the coils.

COMMENTS [1]A spindle or armature is connected to the core and this may be "unguided", that is centralised, by the external attachment point or "guided" where the spindle is controlled by bearings incorporated within the transducer housing; the former eliminates wear of the spindle. [2]The LVDT, used for displacement measurement, is possibly its best application with an extremely wide choice of range, physical size and application. [3]Typical but not limited to those shown. [4]Directly related to the quoted ranges.

SUPPLIERS LUC1; LUC2; PEN; RDP; RSC; SHI.

DL5. Contacting, linear variable differential transformer, gauging

OPERATING PRINCIPLE Although in principle these LVDT devices are as described in DL4, they have been identified here separately because they represent a special class in that their purpose is to provide an electrical equivalent of the mechanical dial gauge.

APPLICATIONS These devices are the electrical equivalent of the dial gauge or micrometer and are used in similar contexts but allowing remote operation. They are particularly useful for precise measurement even when experiencing motion.

Courtesy RDP Group

RANGE LIMITS ±0.013 mm to ±25 mm.

ERROR Nonlinearity: ±0.2% or ±0.05% of full range. Repeatability: ±0.0001 mm.

OVER-RANGE PROTECTION High and provided with a generous travel beyond the normal displacement capability.

SUPPLY VOLTAGE Transducer: 3 V rms at 2.5 kHz. Signal conditioning: ±15 V dc

SENSITIVITY Transducer: 16 mV/mm/V to 250 mV/mm/V.
Signal conditioning: ±10 V full range output.

CROSS SENSITIVITY Extremely low particularly for guided systems.

TEMPERATURE RANGE Typical ambient temperatures.[1]

TEMPERATURE SENSITIVITY Zero: ±0.01% of full range/°C.
Range: ±0.01% of full range/°C.

SIGNAL CONDITIONING Both options of ac energising of the transducer alone and dc energising of a combined transducer and signal conditioning are available.

BENEFITS AND LIMITATIONS The equivalent of mechanical devices but offer electrical outputs giving automatic and rapid measurement with the potential for data handling.

COMMENTS [1]These devices are generally used in a workshop environment and thus the normal operating temperature range will be ambient. However the capability will be generally in line with other LVDT displacement devices. These devices are generally the most precise of the displacement transducers listed in this section.

SUPPLIERS LUC1; LUC2; RDP; RSC.

DL6. Contacting, resistive

OPERATING PRINCIPLE A resistance element of wire or a hybrid of conducting plastic and wire is housed in the body of the transducer.
This body guides a rod which incorporates a "wiper" which slides across the resistance element.
A measurement of the resistance between the slider and either end of the resistance element provides a measure of displacement.[1]

APPLICATIONS Used in a wide range of control engineering applications particularly where a high signal output is required or a larger stroke is essential.

Courtesy RS Components Ltd

RANGE LIMITS 0 – 10 mm to 0 – 2 m.

ERROR Nonlinearity: ±0.1% of full range. Resolution: ±0.02% of full range.

OVER-RANGE PROTECTION Good with some overtravel incorporated.

SUPPLY VOLTAGE Up to 100 V for the longer stroke devices.[2]

SENSITIVITY The full range voltage is coincident with the supply voltage.

CROSS SENSITIVITY Being guided devices there are no transverse sensitivity effects as such.

TEMPERATURE RANGE Operating: –50°C to +200°C.

TEMPERATURE SENSITIVITY Temperature coefficient: ±120 ppm/°C.

SIGNAL CONDITIONING Since the voltage available at full scale is so large it is unlikely that any further signal conditioning will be required.[3]

BENEFITS AND LIMITATIONS As a sliding contact device wear will occur but values of life expectancy of 50×10^6 cycles or more are possible.

COMMENTS [1]The devices can be spring loaded or directly attached and can be operated from the side or by a magnetic ring; where space may be limited. Some devices incorporate two resistance tracks and two mechanically-connected wipers to provide a dual output.
[2]Care must be taken to limit the maximum current through the winding.
[3]However, hybrid devices must be used as voltage dividers and do require a high wiper circuit impedance.

SUPPLIERS PEN; PRE; RSC; SOI; TEC.

DL7. Noncontacting, acoustic

OPERATING PRINCIPLE A piezoelectric crystal is energised so as to transmit bursts of acoustic signals through a gaseous medium, usually air.
The return echo is detected by the same crystal and "the time of flight" of the signal is measured.
This measure of time is proportional to the distance travelled.
Any surface, solid or liquid, can be detected in this manner.

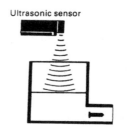

Ultrasonic sensor

APPLICATIONS Clamp-mounted at the detection point, they are highly suited in situations which exploit their long range such as tank liquid levels or tide gauges.

Ultrasonic sensing of liquid level
Courtesy Butterworth-Heinemann Ltd

RANGE LIMITS 0.7 m to 3.5 m.

ERROR Nonlinearity: ±1% of full range.

OVER-RANGE PROTECTION None required.

SUPPLY VOLTAGE AC mains.

SENSITIVITY 1 V dc/m of displacement or 4 – 20 mA for full range output.

CROSS SENSITIVITY Contamination of the pathway of the acoustic signal, such as dust or steam, may result in a dramatic deterioration of performance.[1]

TEMPERATURE RANGE Operating: –20°C to +70°C.

TEMPERATURE SENSITIVITY ±0.06% of full range/°C.[2]

SIGNAL CONDITIONING Supplied systems provide the acoustic drive signal, the return signal detection and possibly alarms including such features as "no transmission pulse".

BENEFITS AND LIMITATIONS Insensitive to the type of material which it detects and will operate over greater distances than any other type listed here.

COMMENTS [1]Although the acoustic transmission is conical in form the system will be so engineered as to measure the earliest part of the return signal, that is the shortest time signal. This may not be the required distance unless proper care is taken.
[2]Acoustic transmission is influenced by temperature of the transmission medium. It is normal to measure the temperature at the transducer head and apply a suitable correction assuming that it is representative of the total path. These quoted values are based upon this assumption.

SUPPLIERS MIC1; MIC2.

Displacement, linear

DL8. Noncontacting, capacitive

OPERATING PRINCIPLE A transducer capacitor plate is energised by a fixed frequency current.
As its plate approaches the desired surface of the target the transducer will experience a change of reactance resulting in a change of voltage, proportional to the separation distance.
In order to concentrate the area under measurement, a shield or guard ring is located around the capacitor plate and this further improves the linearity.[1]

APPLICATIONS Clamp-fitted, they are used for position control purposes on flat surfaces such as machine tool positioning or where nonmetallic surfaces are involved.

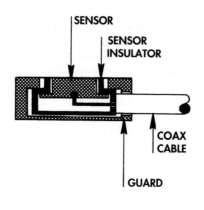

Courtesy Anley Controls Ltd

RANGE LIMITS 0 – 0.05 mm to 0 – 26 mm is typical.

ERROR Nonlinearity: ±0.1% of full range.

OVER-RANGE PROTECTION Good. However care must be taken to avoid driving the transducer into the target surface.

SUPPLY VOLTAGE Signal conditioning: 110/240 V ac mains. Transducer: low voltage, 20 kHz at constant current.

SENSITIVITY 0 – 10 V for full range.

CROSS SENSITIVITY Good. They incorporate both a ground ring and shield ring electrodes to provide protection from external influences.[2]

TEMPERATURE RANGE Operating: –50°C to +150°C.[3]

TEMPERATURE SENSITIVITY Zero: ±0.03 mm/°C to ±0.17 mm/°C.[4]
Range: ±0.0003% of full range/°C to ±0.003% of full range/ °C.[4]

SIGNAL CONDITIONING Part of a complete system with a wide range of sophistication.

BENEFITS AND LIMITATIONS Good long-term stability, typical of capacitive devices but the area coverage may inhibit their use for some applications.

COMMENTS [1]Can be used for metals and insulators and both static and vibratory measurement, (up to 6 kHz). An ac-coupled amplifier will eliminate any dc component.
[2]Since these devices measure over an area, the output will be a representation of that area. (Probe sizes: lowest range 3 mm dia, highest range 60 mm dia).
[3]This assumes a relative humidity of less than 95%, noncondensing. Thus care must be taken below 0°C.
[4]Lowest to highest range.

SUPPLIERS AUT; MIC1; MIC2; RDP.

Displacement, linear

DL9. Noncontacting, eddy current

OPERATING PRINCIPLE A high frequency current energises a coil whose axis is vertical to the measurement plane. This generates an electromagnetic field which when located near to an electrically-conducting surface causes eddy currents to be induced within the coil.

These eddy currents are detected by measuring the change of coil impedance; caused by the absorption of some of the energy within a resonant circuit.

The energy change is directly related to the distance from the conducting surface.[1]

APPLICATIONS Clamp-mounted, they are used for monitoring displacement of flat metal surfaces such as machine tools or other moving surfaces.

Courtesy Lucas Schaevitz Ltd

RANGE LIMITS 0 – 0.5 mm to 0 – 80 mm.

ERROR Nonlinearity: ±0.2% of full range. Resolution: ±0.01% of full range.

OVER-RANGE PROTECTION Good. However care must be taken to avoid driving the transducer into the target surface.

SUPPLY VOLTAGE 5 V dc, ±15 V dc or mains AC.

SENSITIVITY 10 V for full range, with signal conditioning.

CROSS SENSITIVITY Good. They incorporate shield ring electrodes to provide protection from external influences.[2]

TEMPERATURE RANGE Compensated: +10°C to +90°C. Operating: –50°C to +350°C.

TEMPERATURE SENSITIVITY Compensated: ±0.02% of full range/°C.

SIGNAL CONDITIONING This is part of an integrated system with varying degrees of sophistication.

BENEFITS AND LIMITATIONS Good long-term stability typical of inductive devices but are confined to metal component usage.

COMMENTS They can measure vibratory motion. The use of an ac-coupled amplifier will eliminate the dc component of displacement. [1]Used for both ferromagnetic and nonferromagnetic metals. To concentrate the magnetic field a "shield ring" may be incorporated around the sensor head. [2]Since these devices measure over an area, the output will be an average distance from that area. (Effective diameter: unshielded, ×3 of sensor dia; shielded devices ×1.5 of sensor dia.)

SUPPLIERS HBM1; HBM2; MIC1; MIC2; RAL.

DL10. Noncontacting, inductive

OPERATING PRINCIPLE An inductance has its axis placed vertical to the plane of sensing.

As it approaches a target of magnetic or ferrous material the inductive value is changed.

By making the coil part of a resonant circuit, changes in the circuit frequency gives a direct measure of the displacement.

APPLICATIONS Clamp-mounted, they are used for simple control functions, such as for safety systems or "over-run" applications.

Courtesy RDP Group

RANGE LIMITS 0 – 0.8 mm to 0 – 35 mm.

ERROR Nonlinearity: ±1.0% of full range. Resolution: ±0.1% of full range.

OVER-RANGE PROTECTION Good. However care must be taken to avoid driving the transducer into the target surface.

SUPPLY VOLTAGE Depends upon signal conditioning but both low voltage dc or ac mains systems are available.

SENSITIVITY 10 V for full range, with signal conditioning.

CROSS SENSITIVITY Good, but will be influenced by stray magnetic fields.

TEMPERATURE RANGE –50°C to + 150°C.

TEMPERATURE SENSITIVITY ±0.02% of full range/°C.

SIGNAL CONDITIONING This is part of an integrated system with varying degrees of sophistication.

BENEFITS AND LIMITATIONS Well-suited for ferrous metal detection.

COMMENTS Perhaps less popular than the eddy current devices described at DL9, but in suitable applications they are cost effective.

SUPPLIERS HBM1; HBM2; MIC1; MIC2; RDP.

Extension

The measurement of the relative movement between two points of the same body is an essential requirement to the study of the behaviour of materials. The evolution of **extensometers**, starting around the turn of the 20th century, led to a better understanding of the physical characteristics of metals. As techniques were developed and improved, the determination of the elastic behaviour, yield and ultimate failure were subjects of considerable study.

A remarkable range of mechanical devices were conceived including mechanical linkage, optical and pneumatic devices, some of which even included autographic recording within their capability.

Extensometers may be classified under two types of transducer, namely those devices which measure the extension from some remote position and those which are self-contained on the specimen. In general, the advantage of the former was and remains that by using remote optical systems much greater resolution is achievable. Self-contained methods provide a simple system capable of use in a wide variety of circumstances which overcomes the problems caused by bulk movement of the body producing movement in the measurement reference points. Novel techniques are often used to overcome this particular problem.

The search for understanding of the physical properties of materials continues today particularly for high temperature and cryogenic applications. Indeed, the routine testing of materials as part of modern quality control is now well established and would appear to be on the increase.

Extensometers have other roles to play and we find it is with these other roles that the occasional misnaming of them arises. In the various devices listed in this section some are used for load, stress or strain determination. However it must be realised that they are all extensometers at least in the definition given earlier. That there is confusion is in no doubt. Whilst preparing this book, the author noted one manufacturer's multi-lingual text referring to a "strain gauge" in one language and "extensometer" in another. (As early as 1928 there was a paper written identifying the vibrating wire or acoustic gauge as a "strain gauge".) The matter becomes worse as other misnomers arise. Extensometers are sold as "stress" gauges and load cells, neither of which are true unless a considerable amount of other information and/or calibration is included with the measurement.

Modern extensometers are easy to use but two things must be remembered. Extensometers are generally designed to measure the "straight line" distance between the two locations under consideration and during the measurement these locations or gauge points must remain unchanged during the measurement cycle.

When used for the determination of strain the attachment points must be known with precision and further if a test specimen is likely to experience bending then at least two and preferably three extensometers may be required simultaneously so that the bending can be identified and eliminated from any subsequent calculation.

Like their original counterparts, modern optical extensometers have the potential to provide the highest order of accuracy but electrical devices have an important role and have generally replaced many of the earlier mechanical transducers.

The size, shape and method of attachment of the extensometers described here vary considerably. However they may be classified both by their method of attachment and the manner in which the movement is translated into deformation. These are identified here as:

Type 1. These are shaped like a tuning fork or a "U" with the whole mounted perpendicular to the surface by the protruding prongs; the ends of which may include knife edges to provide a precision location.

This type has a further subdivision into **clip gauges** which are mounted directly to and are supported by the material under test by the use of springs or spring-loaded clips; and another type which, being bulkier, has a considerable part of their mass supported externally. These latter devices are of much greater complexity than the clip gauge, particularly in the signal conditioning, and may be considered as part of a complete and often complex testing system. Thus such types are not included here. The minimum and maximum width, thickness or diameter of specimens and the nature of the designs of clip gauges imposes restrictions on their proper use.

Type 2. These lay parallel and close to the surface of interest and are attached by some means such as bolts, welding or by bonding with a suitable adhesive.

Reference will be made to these two types in those listed but potential users will need to discuss in detail with manufacturers the best methods of attachment. Type 1 are designed almost exclusively as true extensometers and used for material behaviour studies whilst Type 2 have a much more varied role including strain, stress and load measurement.

Each device will have some designed set length over which it will operate together with a measurement excursion. It normally will operate both in the compressive and tensile mode but the majority of testing is undertaken in tension. Where the values are quoted as "±" values it follows that for single-direction loading the total excursion will be numerically twice the quoted value.

Extensometers are used both in the static and dynamic mode. Their frequency response is dependent upon the design and values are quoted where available.

Extensometry is a skilled craft and some equipments are of great sophistication, specially designed to allow repeated routine testing specific to a given requirement. Certainly the measurement of extension requires care and the manufacturers should be consulted prior to embarking upon a programme of measurement.

Although there is the usual range of sensing elements, individual designs will vary considerably. Thus the devices described tend to be from a single manufacturer. No special merit is claimed for those chosen other than they are from reputable manufacturers with long experience.

Calibration requires an adequate accuracy rig for the Type 1 devices but, depending upon the desired performance, it is possible to undertake in-house checks with high quality workshop measuring equipment and a certain amount of ingenuity. On the other hand Type 2 equipment is usually purchased with the extension/output relationship specified by the manufacturer. However some testing may be necessary to ensure that the chosen method of attachment is consistent and the effective gauge length is established.

NAMAS Accredited Laboratories: 0019, 0052, 0066, 0086, 0090, 0135, 0167, 0188, 0252, 0259 and 0306.

Contents: Extension

EX1. Bonded resistance strain gauge, type 1

Gauge length: 3 mm to 50 mm. Excursion range: ±1 mm to ±4 mm

EX2. Capacitive, type 1

Gauge length: From 5 mm – 10 mm to 5 mm – 50 mm. Excursion range: As for gauge length.

EX3. Capacitive, type 2

Gauge length: 20 mm. Excursion range: 0.2 mm.

EX4. Magnetic probe, type 2

Resolution: ±0.1 mm.

EX5. Vibrating strip, type 2

Gauge length: 40 mm. Excursion range: ±10 mm.

EX6. Vibrating wire, type 2

Gauge length: various. Excursion range: ±3000 micro strain.

Extension

EX1. Bonded resistance strain gauge, type 1

OPERATING PRINCIPLE Two arms project from the transducer body and are placed in intimate contact with the specimen via knife edges, specimen contact being maintained with spring clips.

At the other end, resistance strain gauges are bonded on one or both arms of a bending transducer which links them.

Movement of the knife edges results in a strain response detected by the strain gauge bridge whose output is a direct measure of the extension.

APPLICATIONS A good general-purpose device commonly used for research applications into the mechanical properties of materials, particularly under low temperature conditions.

Courtesy Active Load Ltd

RANGE LIMITS Gauge length: 3 mm to 50 mm.[1] Excursion: ±1 mm to ±4 mm. Frequency range: 0 to 150 Hz.

ERROR Nonlinearity: ±0.15% of full range. Hysteresis: ±0.1% of full range.

OVER-RANGE PROTECTION Specimen failure is most likely to occur first. It may be necessary to protect the device from damage resulting from sudden release.

SUPPLY VOLTAGE 10 V maximum.[2]

SENSITIVITY Up to 3 mV/V of supply.

CROSS SENSITIVITY Insignificant, since any transverse strain signals will be low and is further counteracted by the chosen gauge configuration.

TEMPERATURE RANGE Operating: –265°C to +600°C.[3]

TEMPERATURE SENSITIVITY Zero: ±0.01%/°C of full range. Range: ±0.02%/°C of full range.[4]

SIGNAL CONDITIONING Normal strain gauge signal conditioning or differential amplifiers combined with a stable power supply may be used.

BENEFITS AND LIMITATIONS Lightweight. Allows conventional strain gauge equipment to be used. Useful where a combined strain and extension programme is under way.

COMMENTS [1]Example. Some designs use only one strain gauged arm, with the other arm being of high stiffness to give a greater gauge length adjustment.
[2]Lower voltages may be necessary, between 1 and 3 V, to prevent self-heating and output drift.
[3]At the higher temperatures isolated by the use of special knife edges and water cooling of the transducer.
[4]Also affected by the coefficient of expansion of the specimen.

SUPPLIERS ACT; MEA1; MEA2.

EX2. Capacitive, type 1

OPERATING PRINCIPLE Each of the two prongs of the transducer form parts of a capacitor with the body forming the others.

Deformation of the prongs results in plate separation causing a differential change of capacitance proportional to the end movement of the prongs. Contact with the specimen is maintained by springs or clips.

APPLICATIONS These clip gauges are specifically designed for high temperature applications although they will operate over a wide temperature band.

Courtesy Anley Controls Ltd

RANGE LIMITS Gauge length: From 5 mm – 10 mm to 5 mm – 50 mm. Excursion range: as for gauge length. Frequency range: 0 – 2 Hz.

ERROR Nonlinearity: ±0.5% of full range.

OVER-RANGE PROTECTION Not normally at risk. Care should be taken at the lower temperatures to protect against ice formation on the capacitor plates.

SUPPLY VOLTAGE Low voltage ac.

SENSITIVITY 10 V dc for full range with signal conditioning.

CROSS SENSITIVITY Insignificant. The design will generally preclude transverse sensitivity due to the manner in which deflection is transferred through the contact points.

TEMPERATURE RANGE −55°C to +650°C

TEMPERATURE SENSITIVITY Dependent upon the differential expansion between the test material and the extensometer.

SIGNAL CONDITIONING From a simple system such as a multimeter to one of much greater complexity dependent upon the desired accuracy.

BENEFITS AND LIMITATIONS Capacitance devices provide excellent stability and resistance to high temperature without the need for special cooling facilities.

COMMENTS Capacitive devices will operate in high magnetic or nuclear radiation fields.

SUPPLIERS ANL.

EX3. Capacitive, type 2

OPERATING PRINCIPLE The transducer consists of a metal plate formed into a shallow arc of which the ends are mounted on to a base plate.
On each of these two plates are mounted capacitor plates.
The base plate is attached to the surface under scrutiny by welding and when extension occurs the capacitor plates gap changes.
The resulting change in capacitance is a direct measure of the extension.[1]

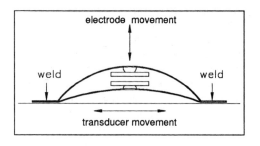

APPLICATIONS Attached by specialised welding techniques, they are used for high temperature research such as steam turbines or nuclear energy plant.

Courtesy Gaeltec Ltd

RANGE LIMITS Gauge length: 20 mm nominal. Excursion range: 0.2 mm, equivalent to 10,000 micro-strain.

ERROR Dependent upon the type of application.[2]

OVER-RANGE PROTECTION Not applicable. Operates over a high strain range generally in excess of strains likely to be experienced.

SUPPLY VOLTAGE Low voltage ac.[3]

SENSITIVITY Gauge factor: $\times 100$.[4]

CROSS SENSITIVITY The transverse sensitivity should be extremely low.

TEMPERATURE RANGE $-269°C$ to $+750°C$.

TEMPERATURE SENSITIVITY Temperature Coefficient: 11 ppm/°C, for ferritic steels and 17 ppm/ °C for stainless steels.[5]

SIGNAL CONDITIONING A suitable capacitance measuring system is required; the degree of sophistication depending upon the use.

BENEFITS AND LIMITATIONS Essentially for strain measurement applications. Below 0°C it will be necessary to ensure that the capacitor plates do not "ice up".

COMMENTS [1]The shape ensures a mechanical magnified movement of the capacitor plates, compared to the extension.
[2]Somewhat nonlinear with the output related to the tangent of the angle between the base and arc plate at their junction. The overall performance is influenced by the mounting and subsequent use.
[3] At "zero" the capacitance of the plates is between 0.6 and 1 pF.
[4]For an explanation of gauge factor see "Strain".
[5]Temperature drift: 0.02 micro-strain/hour (possible).

SUPPLIERS GAE.

Extension

EX4. Magnetic probe, type 2

OPERATING PRINCIPLE Ring-shaped magnets sliding over a central access tube are lowered into a bore hole. The magnets are then located positively (grouting or leaf springs) and a probe carrying one or more reed switches is lowered into the access hole using a steel measuring tape. By detecting the operation of the reed switch as it comes into proximity of the magnet the distance from a convenient datum is measured by the tape.[1]

APPLICATIONS These devices are used for earth movements such as rock slippage and are used extensively by the civil engineering industry.

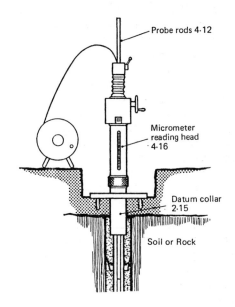

Courtesy Soil Instruments Ltd

RANGE LIMITS Use in bore holes of up to 100 m long.

ERROR Resolution of position: ± 0.1 mm[2]

OVER-RANGE PROTECTION None required.

SUPPLY VOLTAGE Any low voltage supply to the reed switch.

SENSITIVITY ± 0.1 mm.[2]

CROSS SENSITIVITY Any change of the tape path which is not a reflection of movement of the magnets will give rise to error.

TEMPERATURE RANGE Normal ambient temperatures but should operate over a greater range.

TEMPERATURE SENSITIVITY A function of possible expansion of the tape or rod.

SIGNAL CONDITIONING The output is fed to a buzzer or any other desired simple indicator.

BENEFITS AND LIMITATIONS Simple and precise for the industry for which it is built.

COMMENTS Although this device is described for earth studies it can be adapted for other long-based measurements.
[1]Alternatively, for holes which may be inclined or for greater accuracy a steel bar may replace the tape and a micrometer or displacement transducer may be used for final measurement.
[2]Subject to the quality of the chosen form of measurement.

SUPPLIERS SOI.

Extension.

EX5. Vibrating strip, type 2

OPERATING PRINCIPLE A steel strip, shaped in a shallow "S", is inclined to and clamped down at both ends on a high tensile steel ring.[1]
Diametral deformation of the ring will result in a change of shape resulting in a change of natural frequency.
A coil magnet pulses the strip into vibration and then detects the generated ac signal as the vibration decays.
The frequency of the output is a measure of the extension.

APPLICATIONS Attached by welding feet incorporated in the design, they are used widely on structures such as off-shore rigs.
Particularly well-suited for hostile environments.

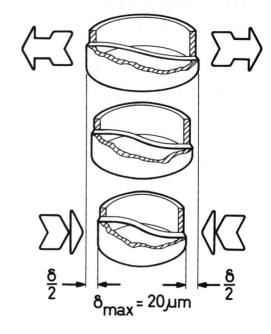

$\frac{\delta}{2}$ $\delta_{max} = 20 \mu m$ $\frac{\delta}{2}$

Courtesy Ranco Controls Ltd

RANGE LIMITS Gauge length: 40 mm only. Excursion range: ±10 mm.

ERROR Nonlinearity, hysteresis and repeatability: ±1% of full range.

OVER-RANGE PROTECTION x10 of full range.

SUPPLY VOLTAGE Self-generating output but does require an electrical pulse to the coil for vibration initiation.

SENSITIVITY A frequency change of 16 to 20 Hz/mm.

CROSS SENSITIVITY Subject to the extension being transferred through the feet alone the transverse sensitivity should be insignificant.

TEMPERATURE RANGE −30°C to +80°C.

TEMPERATURE SENSITIVITY Range: ±0.01% of full range/°C.[2]

SIGNAL CONDITIONING Signal conditioning requires a suitable pulse supply and a frequency counter.[3]

BENEFITS AND LIMITATIONS Rugged and with good long-term stability. Lead resistance and thermovoltage effects have little or no effect upon the frequency output.

COMMENTS [1]In this condition the strip is unstrained.
[2]Based upon attachment to steels with a coefficient of expansion of 11.5×10^{-6} of full range/°C.
[3]The frequency range is ±120 Hz about a centre frequency of 1600 Hz. The devices do require a degree of force, in the order of 60 N, to deform them over the full range. However for most applications this will normally be of little import.

SUPPLIERS APP; DEU; RAN1; RAN2.

Extension

EX6. Vibrating wire, type 2

OPERATING PRINCIPLE A vibrating wire device operates much like the string of a musical instrument.
When plucked a pretensioned wire will vibrate at its natural frequency.[1]
If the wire is then subjected to a change of tension the frequency will change. The wire is "plucked" by a pulse of dc voltage applied to an electromagnet, centrally located, and the resulting vibratory signal is picked up by the same or a second electromagnet.

APPLICATIONS Attached by bonding, bolting or welding, they are in common use in the civil engineering industry for both concrete and steel measurements, either surface mounted or embedded.[2]

Courtesy Soil Instruments Ltd

RANGE LIMITS Strain range: ±3000 micro-strain. Frequency range: 500 to 1100 Hz.

OVER-RANGE PROTECTION Usually rather limited but some devices allow setting of the wire tension after installation thus accommodating any preset extension.

ERROR Nonlinearity and hysteresis: ±0.5% of full range.[3]

SUPPLY VOLTAGE A self-generating output but requires an initial pulse to the coil for vibration initiation.

SENSITIVITY Between 0.5 and 1 micro-strain.

CROSS SENSITIVITY Insignificant since the device always attempts to measure the "straight line" distance between the mounting points.

TEMPERATURE RANGE −40°C to +65°C

TEMPERATURE SENSITIVITY Low when mounted on steel.[4]

SIGNAL CONDITIONING Special-purpose equipment can be obtained to provide the excitation pulse, frequency and a direct read out.

BENEFITS AND LIMITATIONS For continually changing extension it may be necessary to measure the period rather than frequency.

COMMENTS [1] The relationship is: Frequency $= (tg)^{0.5}/2L_w\sigma$ where t = tension, g = gravity, L_w = wire length and σ = density of wire.
[2] Used for both surface mounting and embodiment into materials such as concrete.
[3] After signal conditioning linearisation.
[4] For other materials the transducers may incorporate a temperature monitoring sensor.

SUPPLIERS ROC; SOI.

Position and proximity

Within the chosen definition discussed earlier, the measurement of position needs to satisfy just a single activity, to operate as a fiducial indicator. It operates essentially as a switch when a desired precise position has been reached. It is commonly used as part of a length or shape measuring machine, whereby once triggered some other, more complex, system is used to undertake the absolute measurement of length.

Proximity devices are also used to identify that a location has been reached by initiating a switching action. The difference is in the degree of precision. Examples of proximity switching include the control of liquid in a container to avoid overfilling and, perhaps the most common, to indicate a simple status such as a door closed or open. Many proximity devices are used for counting events such as the passage of components along a conveyor belt.

The degree of precision for both proximity and position will be related to the need and can vary between millimetres and nanometres or better. Often the position relates to a single direction but there are devices which will accommodate two or more axes.

Of course any electromechanical switch must operate as an indicator but, except for the magnetic reed switch, only those devices which are purpose-built for the task are described.

Unlike any other device listed within this book, hysteresis of a position/proximity transducer is not necessarily an undesirable feature. The degree of hysteresis which can be tolerated or indeed desired will be determined by the requirement. For precision measurement it will be low whereas for maintaining a level in a tank by using a pump, a high value of hysteresis may be desirable to reduce the number of pump start-ups.

Calibration must be achieved by repetitive testing against some fixed position measurement. It is important that during this performance testing, checks are made to establish likely detrimental influences such as the introduction of liquids on surfaces; indeed any test which will simulate possible operating conditions.

NAMAS Accredited Laboratories.

There are a very wide range of laboratories which provide some form of calibration for position and proximity. To identify them individually together with their particular speciality would serve little purpose and possibly confuse since the methods adopted to provide the calibration will depend upon individual requirements. The reader is advised to contact NAMAS for advice.

Contents: Position and Proximity

Note. The range values quoted below describe only the general limits of detection.

PO1. Liquid level, piezo-electric

±1 mm.

PO2. Position switch, electromechanical

$0 - 0.35\ \mu$m.

PO3. Position switch, force balance

$0 - 0.25\ \mu$m.

PO4. Proximity switch, capacitive

4 mm to 65 mm.

PO5. Proximity switch, eddy current

0.6 mm to 18 mm.

PO6. Proximity switch, inductive

2 mm to 15 mm.

PO7. Proximity switch, magnetic reed

±1 mm.

Position and proximity

PO1. Liquid level, piezo-electric

OPERATING PRINCIPLE At the sensing position two piezo-electric crystals face each other.

One, the transmitter, sends a continuous signal to the other receiver crystal.

The matching of the crystals with their environment is chosen so that only when a liquid is present in between is there a significantly large signal transmitted across the gap.

The received signal operates a switch used in the "normally on" or "normally off" modes.

APPLICATIONS Essentially for liquid level control in tanks and vats.

May also be used for water level detection in drainage systems.

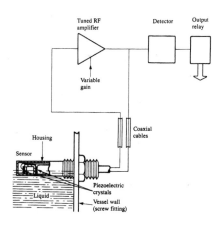

Ultrasonic liquid level sensor circuit
Courtesy Butterworth-Heinemann Ltd

RANGE LIMITS ±1 mm	
ERROR Hysteresis: ±1 mm	

OVER-RANGE PROTECTION They may operate in the "drowned" position thus this device cannot be over-ranged or over loaded.[1]

SUPPLY VOLTAGE 24 V dc or mains ac.

SENSITIVITY A dramatic change of output during transition, gas to liquid.

CROSS SENSITIVITY Not applicable but may be sensitive to surface ripples.

TEMPERATURE RANGE Operating: –40°C to + 150°C (immersed part).[2]

TEMPERATURE SENSITIVITY Insignificant.

SIGNAL CONDITIONING Incorporated within the device and includes an output switch of sufficient capacity to operate many types of electrical load.

BENEFITS AND LIMITATIONS No moving parts and suitable for a wide choice of liquids. May not be suitable for inflammable or corrosive liquids.

COMMENTS An elegant solution to a traditional problem.
[1]The maximum pressure under which the piezo-electric crystals will operate can be as high as 7 MPa dependent upon the ambient temperature.
[2]With a lower maximum temperature for the enclosed electrical system.

SUPPLIERS KDG; PLA.

Position and proximity

PO2. Position switch, electromechanical

OPERATING PRINCIPLE Consists of a stylus which, through its spherical- or disc-shaped tip, makes contact with the surface being measured with its other end attached to a group of electrical switches – typically three minimum to allow multi-axis measurement.
When the stylus contacts the surface, its resulting displacement causes one or more of the switches to change status.
The stylus may have a multiplicity of arms to accommodate the multiple axis measurement function.[1]

APPLICATIONS With plug-in mounting, these devices form part of precision, laboratory-based measuring machines.

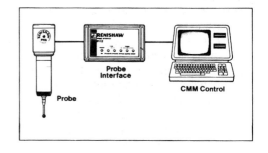

Courtesy Renishaw Metrology Ltd

RANGE LIMITS Not applicable; these are touch sensors.

ERROR 0.35 μm or better in all axes.

OVER-RANGE PROTECTION Provided by allowing the stylus to have considerable spring-loaded over-range movement.[2]

SUPPLY VOLTAGE To suit requirements but without damaging the switch contacts.

SENSITIVITY Dependent upon use. Sensitivity is likely to be different for some axes.[3]

CROSS SENSITIVITY Insignificant but dictated by the true shape of the stylus tip compared to its assumed one.

TEMPERATURE RANGE Usually confined to temperatures close to 20°C.

TEMPERATURE SENSITIVITY Low except for the stylus temperature sensitivity along its length.

SIGNAL CONDITIONING The probe forms part of a servo measuring system. Switch signal transmission can be by cable or indirect methods.

BENEFITS AND LIMITATIONS Simple and reliable in operation with a very wide range of probe/stylus options.

COMMENTS [1]For precision, the stylus needs a minimum of deformation at the moment of switching, yet must provide an override condition whereby the stylus tip may continue its motion after contact. Hence spring loading is used to accommodate the additional travel.
[2]Dependent on the capability of the measuring machine to stop or retrace its movement before full over-range is reached.
[3]Influenced by the contact force and elasticity of the stylus and overload spring at the moment of contact.

SUPPLIERS REN.

Position and proximity

PO3. Position switch, force balance

OPERATING PRINCIPLE These devices consist of a stylus which, through its spherical or disc-shaped tip, makes contact with the surface being measured.

Its other end is attached to a strain gauge force balance such that when the stylus comes into contact with the surface the generated force or moment results in detection by the balance.[1] The electronic control system detects this change of output and provides an electrical status change.

APPLICATIONS With plug-in mounting, these devices form part of precision, laboratory-based measuring machines.

Courtesy Rensihaw Metrology Ltd

RANGE LIMITS Not applicable; these are touch sensors.

ERROR This will depend upon a number of factors including the stylus length, its stiffness, the velocity of motion of the probe and any extraneous signal noise from the force balance due to local vibration. The performance will lie within the error band which is typically 0.25 μm or better in all axes.

OVER-RANGE PROTECTION Provided by allowing the stylus to have considerable spring-loaded over-range movement.[2]

SUPPLY VOLTAGE Dependent upon the stability requirements of strain gauge devices.

SENSITIVITY Very sensitive to force.[3]

CROSS SENSITIVITY Insignificant but dictated by the true shape of the stylus tip compared to its assumed one.

TEMPERATURE RANGE Usually confined to temperatures close to 20°C.

TEMPERATURE SENSITIVITY Low except for the stylus temperature sensitivity along its length.

SIGNAL CONDITIONING A specialised system to provide the "switching".

BENEFITS AND LIMITATIONS More sensitive than its mechanical counterpart but more complex.

COMMENTS [1]For precision, the stylus needs a minimum of deformation at the moment of switching, yet must provide an override condition whereby the stylus tip may continue its motion after contact. Hence additional spring loading is used to accommodate the additional travel. [2]Dependent on the capability of the measuring machine to stop or retrace its movement before full over-range is reached. [3]The sensitivity is controlled by the minimum detectable electrical signal change; influenced by the force balance stiffness and electrical noise.

SUPPLIERS REN.

Position and proximity

PO4. Proximity switch, capacitive

OPERATING PRINCIPLE These devices sense the change of the electrostatic charge on a capacitance plate as the air gap between the sensor and the surface of interest come into close proximity.
Once this change reaches some suitable value the proximity may be identified.
They are capable of sensing non-conducting materials such as wood or plastic as well as ferrous and non-ferrous metals.

APPLICATIONS A wide range of uses such as product and batch counting; particularly useful for non-metallic items such as cardboard outer cartons.

Courtesy RS Components Ltd

RANGE LIMITS 4 mm to 65 mm.[1,2]

ERROR Unimportant but influenced by the material it is detecting.

OVER-RANGE PROTECTION Other than the possibility of driving the sensor into the target, no protection is required.

SUPPLY VOLTAGE 8 – 30 V dc and 80 – 250 V ac.

SENSITIVITY Switching currents from 0.2 to 5 A are typical.

CROSS SENSITIVITY Care must be taken to ensure that the dielectric property (air gap) is not modified in any significant way.

TEMPERATURE RANGE Operating: –55°C to +85°C.

TEMPERATURE SENSITIVITY Insignificant.[3]

SIGNAL CONDITIONING Supplied as part of the sensor and may include "distance" adjustment and LED indicators to show the "switching" status.

BENEFITS AND LIMITATIONS A simple non-contacting device which allows the monitoring of a wide range of materials.

COMMENTS [1]The higher distances are limited to conducting materials only. Also dependent upon the area of the body being detected.
[2]The response of these devices is up to 250 Hz. (Dependent upon the energising frequency used to charge the capacitor face).
[3]It is important at temperatures below 0°C to ensure there is no formation of ice on the capacitor face.

SUPPLIERS RSC; VER.

PO5. Proximity switch, eddy current

OPERATING PRINCIPLE A coil is energised by an alternating constant current. The resulting electromagnetic field generates eddy currents in the target material resulting in a change of impedance of the coil and hence the voltage amplitude of the input current.

Thus any distortion of the magnetic field caused by an external influence such as the near proximity of a metal, ferrous or non-ferrous, will introduce a change in that amplitude.

APPLICATIONS Suitable for a wide range of location tasks and particularly well adapted for use in fluids such as water or oil.

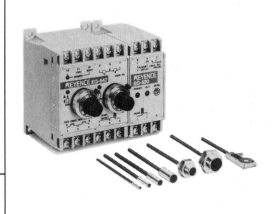

Courtesy Ralcom Automation Ltd

RANGE LIMITS 0.6 mm to 18 mm[1,2]

OVER-RANGE PROTECTION Other than the possibility of driving the sensor into the target no protection is required.

ERROR Tolerance: ±10% tolerance. (Typical). Hysteresis: ±10% of detection distance.

SUPPLY VOLTAGE Low voltage, such as 10 to 24 V dc.

SENSITIVITY Up to 200 mA output current from the signal conditioning.[3]

CROSS SENSITIVITY Detection is over a small area and not a single point. On small targets problems may arise due to off-centre positioning.

TEMPERATURE RANGE Operating: −25°C to +80°C.

TEMPERATURE SENSITIVITY 0.1% of detection distance/°C.

SIGNAL CONDITIONING Completely self-contained systems with LED indication of switch status.

BENEFITS AND LIMITATIONS Best operation is achieved with ferrous material detection.

COMMENTS [1]Range may be influenced by the target materials, ferrous or non-ferrous, and its presented area.
[2]The frequency response of these devices is up to 1.2 kHz.
[3]Devices can be operated as switches which are normally on or normally off.

SUPPLIERS RAL.

Position and proximity

PO6. Proximity switch, inductive

OPERATING PRINCIPLE An inductive coil is wound around a core. When one end of this core is placed near to a metal object there is a change in inductance. This change is measured, using a resonant circuit, and at some preset level the output is used to trigger a switch change.

These devices should not be confused with eddy current devices identified at PO5.

APPLICATIONS Various applications such as aperture status, (open or closed).
Particularly well adapted for use in fluids such as water or oils.

Courtesy RS Components Ltd

RANGE LIMITS 2 mm – 15 mm (ferrous materials).[1,2]	
ERROR See Note 3 below.	
OVER-RANGE PROTECTION Other than the possibility of driving the sensor into the target no protection is required.	
SUPPLY VOLTAGE 10 to 36 V dc or 20 to 250 V ac.	
SENSITIVITY Switched electrical current from 100 mA to 500 mA.	
CROSS SENSITIVITY Detection is over a small area and not a single point. On small targets problems may arise due to "off centre" positioning.	
TEMPERATURE RANGE Operating: –25 degrees C to +85 degrees C.	
TEMPERATURE SENSITIVITY Insignificant in terms of general performance.	
SIGNAL CONDITIONING Usually part of the complete sensor.	
BENEFITS AND LIMITATIONS Restricted to metals. Significantly better with ferrous material detection.	

COMMENTS [1]Values may be decreased to 30% of this range for other metals. The size and position of the presented area will also affect the performance.
[2]The frequency response of these devices is greater than 2 kHz subject to the signal conditioning.
[3]Positional error depends upon the metal, its relative position to the sensor, target area and the signal conditioning and little information is available. However they are likely to have a lower quality of performance than eddy current devices.

SUPPLIERS APP; DEU; RAN1; RAN2; RDP; RSC; VER.

Position and proximity

PO7. Proximity switch, magnetic reed

OPERATING PRINCIPLE Two metal strips are each rigidly fixed at one end and are placed in close location one to the other at their free ends to which are mounted the electrical contacts. The unit is sealed in an evacuated glass surround.
The introduction of a magnet close to the free ends will result in the two contacts touching and hence completing an electrical path.
A permanent or electromagnet may be used.

APPLICATIONS A popular device with many industrial and security uses such as door closure status as in burglar alarms, also rotor-type transducers described elsewhere in this book.

Courtesy RS Components Ltd

RANGE LIMITS ±1 mm of position when the magnet approaches from a single side.[1]	
ERROR See Note 2 below.	
OVER-RANGE PROTECTION These are extremely robust but care must be taken not to exceed the current rating of the switch contacts.	
SUPPLY VOLTAGE Voltage: up to 700 V dc (breakdown). Current: up to 2 A, ac or dc.	
SENSITIVITY Good response to repeated switching.	
CROSS SENSITIVITY Insignificant but see Note 2 below.	
TEMPERATURE RANGE Normal ambient temperatures.	
TEMPERATURE SENSITIVITY No measurable effect over temperature range.	
SIGNAL CONDITIONING When an electromagnet is used manufacturers often supply information as to the design of the coils.	
BENEFITS AND LIMITATIONS With low contact resistance, typically 100 mΩ, they can accommodate high current levels without excessive voltage drop across the contacts.	
COMMENTS [1]Operating distances are in the order of 10 mm and of release 15 mm between the magnet and reeds. The manner of magnet "approach" has significant influence on the manner of opening and closing of these devices. [2]There may be considerable hysteresis between closure and reopening. Sometimes an additional magnet is used to improve this situation. [3]The operating frequency is determined by their natural frequency (2 kHz); giving closure times of 1 millisecond and release times of 0.2 milliseconds.	
SUPPLIERS RSC.	

Strain

For electrical transducers, strain is generally a local measurement upon the surface of the material under study, although occasionally they may be embedded into nonferrous materials such as a glass fibre matrix or concrete. The comments which follow are based upon the surface application technique. Strain in solids is of two forms, linear and shear. Only the former can be measured directly with the latter being determined by the measurement of the principal strains resulting from the shear forces.

Strain as such is a nondimensional quantity since it is the ratio of the extension of some part of a solid body when compared to a datum length. Thus a strain-measuring sensor, called a **strain gauge,** should automatically give that ratio independent of its own length, averaging the strain value over its gauge length.

Unfortunately the words "strain gauge" are often misquoted by identifying devices which are truly extensometers of a predetermined length (see Extension section). Although it is possible to derive strain from such devices, on occasion it is the only suitable technique and it must be remembered that the method of attachment is critical to ensure that the base length is both fixed and known. To call extensometers "strain gauges" hides this weakness from the unwary.

The most common device for measuring strain is the **bonded resistance strain gauge.** These devices comprise a metal grid, usually of foil but occasionally of wire, bonded to, but insulated from, the test surface. When the whole is put under stress the resistance of the gauge changes in direct proportion to the resulting strain.

This relationship is $(dR/R)/(dL/L) = C$, where dR is the change in resistance, R is the initial resistance, dR/R is called the electrical strain, dL/L is the mechanical strain and C is a constant, known as the **gauge factor,** the value of which depends on both the type of metal used for the resistive element and to a much lesser extent on the gauge size and is only a true constant at a constant temperature.

The choice of metal used for forming the grids of wire or foil is influenced by the application. Gauges can be made in a range of metals with both positive and negative gauge factors but in practice only a limited number warrant serious consideration. The choice of a particular metal is influenced by factors such as fatigue life, maximum strain range and temperature characteristics. With the more popular gauges the relationship is such that the gauge factor is approximately 2.

The backings, too, are generally restricted to a few materials and here the selection is influenced by ease of bonding, flexibility, strain transfer capability and temperature performance.

Other than for welding, gauges are generally bonded by an adhesive which must, of course, provide a satisfactory bond to both the gauge backing and the parent surface. Bonding can be achieved at both ambient or elevated temperatures. Bonding as such is a skilful technique and courses are run to ensure that only good techniques are adopted.

Gauges are subject to sources of error. These include:

1. Temperature Coefficient of Resistivity (TCR). This is the resistance change experienced by the gauge alone when subjected to a temperature change resulting in a thermally-induced output. Once the gauge is bonded to a surface the thermal expansion characteristics of that material will influence the thermal properties of the strain gauge. By using a Wheatstone bridge circuit with at least two adjacent arms formed from strain gauges, automatic temperature compensation is provided, eliminating much of the unwanted effect.

Another improvement may be accomplished by using "self temperature compensated" gauges. Here the gauge TCR, at a predetermined temperature, is matched but opposed to the parent material's coefficient of expansion.

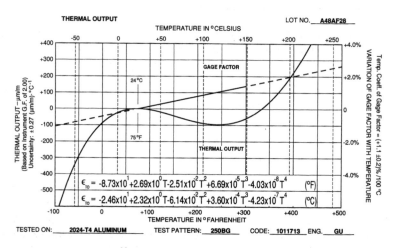

Figure 1.   Typical thermal output and variation of gauge factor with temperature (Courtesy Measurement Group Inc)

2. Temperature Change of Gauge Factor (TCGF). This source of error is exactly as the title implies. For a wide range of applications where the temperature is likely to be close to ambient it is seldom that a correction need be applied but for use over a reasonable temperature excursion, correction is often necessary.

For strain measurement the degree of temperature compensation, if any, will depend upon the aims and purpose of the measurement programme.

For devices which use strain gauges for measurements such as force or pressure, a passive circuit is often included to correct automatically for TCF, TCGF and the changes in the elastic modulus with temperature; usually known as "zero" correction and "range" or "span" correction. (These circuits correct for all of the temperature effects including those due to the bridge circuit wiring.)

3. Gauge Factor. Strain gauges are supplied with the gauge factor value identified. However it is important to realise that this value is obtained by sampling of selected gauges in a unidirectional stress field such as a beam in bending. Thus whilst the gauge experiences a direct strain along its axis it will also experience a further strain in the other plane due to the Poisson ratio effect. This strain will be in the order of 1/3 of the principal strain for metals and if the gauge responds to transverse strain there may be a need to correct for such effects when the gauge operates in a multi-strain field application. The nature of this transverse sensitivity is too complex to develop further here and the reader is recommended to discuss the matter with the gauge suppliers who often will have literature which develops the subject further (see also the Bibliography for further reading material).

Some users prefer to ignore the gauge factor values supplied by the manufacturer and establish values for themselves by local testing on a suitable rig. The claimed benefit is that by bonding gauges to a test beam local to the main activity, it will then make allowance for any effect of local environmental conditions. Although this is a sound approach as a check on the bonding operation, if the values so obtained are significantly different from the original package values then one is left with the dilemma that the error is due to poor bonding and thus the results may be suspect. (That is not to suggest that such techniques be abandoned but one should consider such an approach with proper realisation of its purpose.) While dR remains small compared to R the strain gauge is essentially linear in response. However, for high strain levels, this error source may require some mathematical correction.

4. Circuitry. A strain gauge changes its resistance when under strain and it is possible to measure such change directly. One of the most sensitive methods of detection is by a voltage change in a Wheatstone bridge and nearly all applications, whether for strain measurement or other transducer applications, use this technique. For strain measurement, it is most likely that only one arm of the Wheatstone Bridge will be an active strain gauge and hence there will be a nonlinearity present.

Thus for a single gauge, within a Wheatstone bridge, sensitivity is expressed as:

$$\frac{Vo}{Vi} = \frac{(GF.e)}{4} \times \frac{2}{2+(GF.e)}$$

Where e is the strain, Vo is the output voltage and Vi the supply voltage. Thus a single active arm (gauge factor 2) in a bridge configuration, with a 10 V input, 1 micro-strain (μe) gives a signal value of approximately 5 μV. (Micro-strain is a common manner of expressing strain. 1 micro-strain $= 10^{-6}$ strain). The last portion of the expression is the nonlinear part of the equation and is independent of the linearity of the strain gauge itself. As examples the approximate values of nonlinearity for bonded strain gauges of gauge factor 2 are:

0.1% at 0.1% strain or 1000 μe

10% at 10% strain or 100,000 μe

ie, the nonlinearity expressed as a percentage is approximately equal to the strain expressed as a percentage.

Interestingly, better linearity may be achieved by the use of a constant current source rather than a constant voltage and this is particularly so for semiconductor gauges. However, common practice still results in the constant voltage system being used for the large majority of applications.

Where a single gauge is incorporated into a Wheatstone bridge, it is possible that the gauge will be remote from the other three arms. Under these circumstances problems can arise due to the resistance and temperature characteristics of the connecting wires. Techniques such as "three wire" connections overcome these difficulties. It is important that such problems and solutions are understood and most manufacturers can offer help and literature.

In many applications, strain will be measured under dynamic conditions. Thus gauges will be needed with good fatigue life. Gauges of various levels of fatigue life are available, the value of which is influenced in turn by temperature, strain level and the desired stability of measurement. Manufacturers offer considerable detail in identifying the fatigue performance of their gauges.

Gauges are also made from piezo-resistive material where the gauge factor can be in the order of 150 or more and have positive (P) or negative (N) values. Known as **semiconductor strain gauges,** they are used in transducers where high stiffness or high output is desirable but seldom for strain measurement unless particularly low strain levels are to be measured. Semiconductor gauges should not be considered as a special form of metal gauge since to do so gives rise to invalid comparison. They are unique measurement elements having their own benefits and disadvantages over other options. The measurement elements have the appearance of a single "wire" although some gauges are "U" shaped; these latter have the benefit of the two integral lead-out wires being at the same end of the gauge which often results in more simple wiring paths when completing the resistance bridge network or where space is at a premium. TCR and TCGF effects are also present.

In the text describing the performance of both metal and semiconductor gauges under Error, it will be noticed that the headings of Linearity, Hysteresis and Repeatability are not mentioned. Both types of strain gauge have excellent characteristics; for example, semiconductor material exhibits low hysteresis. However these properties must in the final count depend upon the material under investigation and the quality of the installation. Thus it would be meaningless to quote values for the gauges alone.

In order to establish possible performance it is recommended that appropriate values are gleaned from the wide variety of semiconductor transducers identified in the general text which use these types of sensor. The values quoted there are likely to be the best obtainable.

Some strain gauges are printed onto a ceramic substrate and are known as thick film resistance gauges. They have a gauge factor generally in the region of 10. Although they respond to strain they are particularly sensitive to transverse strain and so are generally only suited for specialised strain measurement applications. They are mentioned here for completeness and are described under pressure transducers, PR11.

NAMAS accredited laboratories: None known.

Contents: strain

ST1. Bonded metal foil resistance strain gauge

Strain range up to 20% strain.

ST2. Bonded metal wire resistance strain gauge

Strain range up to 20% strain.

ST3. Bonded metal wire resistance strain gauge, high temperature

Strain range 0.5% strain.

ST4. Bonded semiconductor resistance strain gauge

Strain range up to 0.5% tensile strain with higher compression values.

ST5. Encapsulated metal resistance strain gauge

Strain range 0.3% strain.

ST6. Weldable metal foil resistance strain gauge

Strain range 0.5% strain.

ST7. Weldable metal wire resistance strain gauge

Strain range 0.3% strain.

Strain

ST1. Bonded metal foil resistance strain gauge

OPERATING PRINCIPLE Comprises a metal foil element formed into a grid and bonded to a substrate of electrical insulating material.
This, in turn, is bonded to the material under test using one of the modern specialised adhesives.
When strained a corresponding electrical resistance change occurs in the metal element.
Available in a wide range of sizes and shapes including multi-gauge patterns giving precise relationships for measurement in a complex strain field.

APPLICATIONS Widely used, both for strain measurement and as the sensing elements of many other types of electrical transducer.

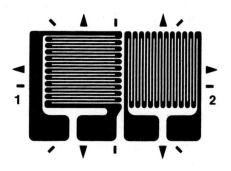

Courtesy Measurements Group (UK) Ltd

RANGE LIMITS General use: 2% strain.[1] Post yield: 20% strain.

ERROR Gauge factor: 0.5%. Resistance: 0.15% to 1.0% of nominal value.[2]

OVER-RANGE PROTECTION All of the strain values quoted are maxima.

SUPPLY VOLTAGE Seldom exceeds 5 V across the gauge.[3]

SENSITIVITY Gauge factor: 2 approximately.[4] Resistance: 0.15% to 0.6% of nominal value.

CROSS SENSITIVITY Foil strain gauges have a response to transverse strain. See also the introduction to this section and in the Bibliography.

TEMPERATURE RANGE −269°C to +370°C.

TEMPERATURE SENSITIVITY TCR and TCGF are dependent upon the type of foil. See Fig. 1 as an example.

SIGNAL CONDITIONING The choice of amplifier depends upon the nature of the strain measurement requirement. Both ac and dc energisation may be used.

BENEFITS AND LIMITATIONS The foil resistance strain gauge is the most common way of measuring strain in a direct manner.

COMMENTS [1]With a wide choice between these two values.
[2]Resistance range: 120 Ω and 350 Ω are the most common but higher values are available, especially for transducer applications.
[3]Dependent upon the heat dissipation capability of the installation, determined by the total wattage, the surface area of the gauge element and the heat dissipation capability of the parent surface.
[4]1 μe can be measured readily in a properly engineered installation.

SUPPLIERS HBM1; HBM2; MEA1; MEA2; NOB; RSC; TEH; TIN.

Strain

ST2. Bonded metal wire resistance strain gauge

OPERATING PRINCIPLE Wire strain gauges comprise a grid of wire bonded to a substrate of electrical insulating material.

The whole is bonded to the material under test using a suitable adhesive. This system will now allow the mechanical strain to induce an electrical resistance change into the wire. Gauges are available in a reasonable range of shapes and sizes including multi gauge patterns giving precise relationships for measurement in a complex strain field.

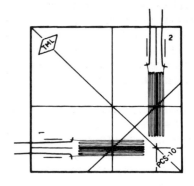

APPLICATIONS Although not as common as foil gauges, having different strain characteristics they are sometimes a preferred choice for strain measurement.

Courtesy Techni Measure

RANGE LIMITS General use: 1% to 3% strain.[1] Post yield: 20% strain.

ERROR Gauge factor: 1.5% is common. Resistance: 1% of nominal value is typical.[2]

OVER-RANGE PROTECTION All of the strain values quoted are maxima.

SUPPLY VOLTAGE Seldom exceeds 5 V across the gauge.[3]

SENSITIVITY Gauge factor: from 2 to 4.5. (Some with negative values).

CROSS SENSITIVITY Wire strain gauges have a response to transverse strain. See also the introduction to this section and in the Bibliography.

TEMPERATURE RANGE −200°C to +300°C.

TEMPERATURE SENSITIVITY Dependent upon type of gauge. Some are identified as self-compensating.

SIGNAL CONDITIONING The choice of amplifier depends upon the nature of the strain measurement requirement. Both ac and dc energisation may be used.

BENEFITS AND LIMITATIONS Wire gauges tend to be used for specialist applications and thus fill that part of the market not readily exploited by their foil counterparts.

COMMENTS [1]With some choice between these two values.
[2]Resistance range: 120 Ω is common but other values ranging from 50 Ω to 1000 Ω are available.
[3]Dependent upon the heat dissipation capability of the installation, the total wattage, and the surface area of the gauge element.
[4]1 μe can be measured readily in a properly engineered installation.

SUPPLIERS NOB; TEC; TIN.

Strain

ST3. Bonded metal wire resistance strain gauge, high temperature

PRINCIPLES OF OPERATION A grid of wire temporarily attached to a substrate of electrical insulating material.
They are bonded to the material under test using a special ceramic-based adhesive and during the bonding process the substrate is removed.
This system will now allow the mechanical strain to induce an electrical resistance change into the wire over a wide temperature range. Gauges are available in a limited range of single gauge size and shape.

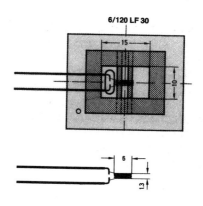

APPLICATIONS Specially developed for high temperature strain such as steam raising plant and turbine research.

Courtesy HBM (UK)

RANGE LIMITS Strain range: 0.5% strain.

ERROR Resistance: 0.7% of nominal value. Resistance: 120 Ω.

OVER-RANGE PROTECTION The strain value quoted is a maximum.

SUPPLY VOLTAGE Up to 2 V bridge supply is recommended.

SENSITIVITY Gauge factor: 4.5.[1]

CROSS SENSITIVITY Wire strain gauges have a response to transverse strain. See also the introduction to this section and in the Bibliography.

TEMPERATURE RANGE Static: –200°C to + 760°C. Dynamic strain: –230°C to + 850°C.

TEMPERATURE SENSITIVITY TCGF: –310 ppm ±5% of nominal value/°C. TCR: not self-compensating.

SIGNAL CONDITIONING Generally the same as for conventional strain gauges . Suitable input/output wiring will be needed to match the temperature requirement.

BENEFITS AND LIMITATIONS These gauges are a specialist device specific to the measurement of strain at extremes of temperature. Bonding is a skilled task.

COMMENTS An application of a general technique applied to an extreme environmental condition. The decision to use these gauges should not be entered into lightly.
[1]The sensitivity of these gauges is in principle as good as conventional gauge installations. However the nature of the application combined with a wide range of uncertainty arising from other aspects of the measurement, suggests a strain level of between 10 μe and 100 μe is a realistic target sensitivity.

SUPPLIERS HBM1; HBM2; TEC.

ST4. Bonded semiconductor resistance strain gauge

OPERATING PRINCIPLE A silicon crystal element is cut so that it exhibits optimum piezo-resistive characteristics; either positive or negative values can be achieved.

When bonded to a material subjected to a change of strain it experiences a significant resistance change.

Normally without an insulating backing since the elements have a surface of sufficient insulating properties which when combined with the adhesive provides the necessary electrical isolation.

Some gauges are supplied on a bondable backing to ease application.

APPLICATIONS Their use is normally for transducer applications such as force, pressure and acceleration

Dual-element self-temperature compensating strain gauge
Courtesy Graham & White Instruments

RANGE LIMITS Strain range: 0.3% to 0.5% tensile; higher in compression.

ERROR Gauge factor: 3% of nominal value. Resistance: 1% of nominal value per set of gauges.[1]

OVER-RANGE PROTECTION All of the strain values quoted are maxima.

SUPPLY VOLTAGE 5 V across the gauge or higher is common.[2]

SENSITIVITY Gauge factor: between +100 and +175 positive types. −100 to −135 for negative types.

CROSS SENSITIVITY Some transverse sensitivity but generally better than foil or wire.

TEMPERATURE RANGE Operating: −200°C to +315°C.

TEMPERATURE SENSITIVITY TCR: +0.05%/°C to +0.8%/°C of nominal value. TCGF: −0.045%/°C to −0.4%/°C of nominal value.

SIGNAL CONDITIONING Conventional strain gauge amplifier systems are suitable. With generally large outputs, low gain amplifiers may be sufficient.

BENEFITS AND LIMITATIONS Their high output for a given strain ensures a high natural frequency for transducer applications.

COMMENTS Gauges are available in a limited range of shapes and sizes but only in single gauge configuration.
[1]Resistance: from 120 Ω to 5000 Ω.
[2]There is considerable benefit in using constant current energisation for these gauges especially for single active arm applications.

SUPPLIERS ENT; KUL; MAY; MIC3; SAN; TEC.

Strain

ST5. Encapsulated metal resistance strain gauge

OPERATING PRINCIPLE This type of strain gauge is a further evolution of the weldable gauge described at ST6. Here, not only is the gauge (or gauges) pre-bonded to a metal or polycarbonate foil but also all wiring, including the lead-out wires, moisture proofing and mechanical protection are completed by the manufacturer. Available in both wire and foil types in limited ranges of size and confined to single gauge or half bridge configurations.

APPLICATIONS These gauges have a wide variety of uses such as on structural steel, offshore structures, ships, concrete and asphalt.

Courtesy HBM (UK)

RANGE LIMITS Strain range: 0.3% strain.

ERROR Gauge factor: 0.5% of nominal value. Resistance: 1% of nominal value.[1]

OVER-RANGE PROTECTION The strain value quoted is a maximum.

SUPPLY VOLTAGE 5 V or less across the gauge.

SENSITIVITY Gauge factor: Slightly lower than the pre-bonded gauges.

CROSS SENSITIVITY Dependent upon the original gauges and the shape and material of the metal backing.

TEMPERATURE RANGE Operating: −200°C to +160°C.

TEMPERATURE SENSITIVITY The TCR and TCGF will be selected for their chosen application.

SIGNAL CONDITIONING Conventional strain gauge amplifiers. Both ac and dc energisation methods are used.

BENEFITS AND LIMITATIONS These gauges, having been presented complete with wiring and protection, do not require any strain gauge installation expertise.

COMMENTS Generally advisable to operate at lower voltages than normal gauge installations, because of the reduction of heat transfer to the base material. May require testing to establish the values. In difficult installation circumstances the attachment of strain gauges by adhesive bonding, wiring and protection will often result in poor final installation. This type of gauge eliminates most if not all of these problems.
[1]Resistance range: 120, 350 and 600 Ω gauges.

SUPPLIERS HBM1; HBM2.

Strain

ST6. Weldable metal foil resistance strain gauge

OPERATING PRINCIPLE Conventional foil strain gauges are bonded to a thin steel foil by conventional means and then the metal foil is spot welded to the parent material.

Gauges are available in a range of sizes and shapes including multi-gauge patterns giving precise relationships for measurement in a complex strain field. Indeed, virtually all types of gauges may be prepared in this way.

APPLICATIONS Offers a rapid installation at difficult locations such as bridges, offshore structures, ships and a wide range of building applications.

Courtesy Measurements Group (UK) Ltd

RANGE LIMITS Strain range: 0.5% strain.

ERROR Gauge factor: 0.5% of the nominal value.
Resistance: 0.4% of the nominal value.[1]

OVER-RANGE PROTECTION The strain value quoted is a maximum.

SUPPLY VOLTAGE 5 V across gauge is the likely maximum.

SENSITIVITY Gauge factor: slightly lower than the pre-bonded gauges.

CROSS SENSITIVITY Dependent upon the original gauges and the shape and material of the metal backing.

TEMPERATURE RANGE Operating: −195°C to + 260°C

TEMPERATURE SENSITIVITY The TCR and TCGF will be selected for their chosen application.

SIGNAL CONDITIONING Conventional strain gauge amplifiers. Both ac and dc energisation is used.

BENEFITS AND LIMITATIONS Requires wiring and protection as for conventional gauges, but it is possible to undertake at least some of this work prior to installation.

COMMENTS Generally advisable to operate at lower voltages than normal gauge installations, because of the reduction of heat transfer to the base material: may require testing to establish the values. There is a limit in the number of types of gauge generally available but some suppliers will build to order for any type of gauge.
[1] Resistance range: 120 Ω or 350 Ω only.

SUPPLIERS HBM1; HBM2; MEA1; MEA2; TEC.

Strain

ST7. Weldable metal wire resistance strain gauge

OPERATING PRINCIPLE These gauges comprise a resistance wire or wires located and held within a metal tube but isolated from it by a refractory material.

The tube, which incorporates two mounting plates along the full length of the gauge, is welded to the parent surface.

Both single arm or half bridge configurations are possible.

In this latter case one arm is strained and the other is used for temperature compensation.[1]

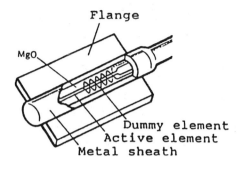

APPLICATIONS Many applications including offshore structures and risers, railway track and concrete embedment gauges.

Courtesy Graham & White Instruments

RANGE LIMITS Strain range: 0.3% strain.

ERROR Gauge factor: 2% of nominal value. Resistance: 0.5 Ω in matched pairs.

OVER-RANGE PROTECTION The strain value quoted is a maximum.

SUPPLY VOLTAGE 5 V across the gauge is typical but may be influenced by the type of installation.

SENSITIVITY Gauge factor: 2.1 for nickel chrome wire, 4 for platinum-tungsten wire (used above 316°C).

CROSS SENSITIVITY This type of gauge has a low transverse sensitivity.

TEMPERATURE RANGE Operating, static: −269°C to +650°C. Operating, dynamic: −269°C to +816°C.

TEMPERATURE SENSITIVITY TCR: 13 ppm of nominal value/°C over compensated range (Ni-Cr). TCGF: −0.018% of nominal value/°C for Ni-Cr and −0.032% of nominal value/°C for Pt-W.

SIGNAL CONDITIONING As for normal strain gauge amplifiers but will depend upon the strain measurement requirement. Both ac and dc energisation may be used.

BENEFITS AND LIMITATIONS Although supplied for attachment to plane surfaces they can be curved around a mandrel to a minimum of 40 mm radius.

COMMENTS [1]Supplied complete with integral lead-out cable. Resistance range: 120 Ω and 350 Ω. Capable of operating in extremes of temperature and pressure.

SUPPLIERS GRA; STR.

Mass (weighing and force)

Possibly the oldest record of establishing the mass of a small item was that left us by the ancient Egyptians. Their burial site drawings show a simple two-scale equi-arm balance being used to weigh precious stones. The word **balance** is from the Latin and simply translated means "two flat plates", a succinct description. From this simple one-to-one comparison evolved the Roman steelyard balance. For rather inexplicable reasons the Danish steelyard balance held sway for some considerable time only to be replaced again by the Roman version. In mechanical terms, such precision balances and steelyards are still in wide usage but are gradually being replaced with electronic devices.

Weighing involves comparing two masses, yet this simple technique does not necessarily lend itself to modern technology nor the need to provide rapid measurements.

There is a dilemma in scientific applications: when a spring device is used properly the unit is the Newton whereas mass comparison requires units of the gram. Thus the measurement unit should be related to the method adopted, force or mass, which for many practical applications would result in total confusion. However force devices are used for most weighing applications and indeed some of the transducers described under Force are applied in this way.

Most weighing activity is linked to statutory regulations and the force of law and thus its general performance is controlled to high standards. In the ultimate, performance quality can be established only by applying known masses to the device and making a direct comparison. Further, unlike force devices, it is an essential requirement that performance is related to the applied weight, at least over some reasonable range, and not expressed as a percentage of the range. Indeed perhaps this difference is a convenient identifier between weighing and force detection transducers when making a choice of equipment suitability. However in a non-ideal world, the practical distinction between weighing and force determination will remain one of convenience and accepted practice.

Calibration checks may be made using any suitable stable mass: preferably one which is identified as traceable to National or International Standards. When standards are not available, coins in good condition are of use, at least for small weights. For example in Britain a mint twenty pence piece weighs 5 grams to a high degree of precision.

NAMAS Accredited Laboratories: 0180, 0134, 0260 and 0304.

Contents: Mass, weighing

WE1. Balance, electronic

3 g to 60 kg.

WE2. Balance, weight comparator

5 g to 60 kg.

WE3. Load cell, industrial platform weighing

500 g to 50,000 kg.

Mass, weighing

WE1. Balance, electronic

OPERATING PRINCIPLE Essentially a servo device comprising an electrically energised inductive coil located around a magnetic load-carrying core.
When load is applied, causing movement of the core, a detector senses this positional change and induces a change in the coil current sufficient to restore the core to its original position.[1]
For the larger ranges, the load carrying is achieved through a mechanical lever system.

APPLICATIONS Used for a wide range of precision measurements where the convenience of use outperforms mechanical balances. Particularly suited for mobile applications. Used for mass calibration.

Courtesy GEC Avery Ltd

RANGE LIMITS 0 – 3 g to 0 – 60 kg.

ERROR Non linearity: ±0.0002 g (110 g). ±0.0015 g (500 g).[2]
Repeatability: ±0.0001 g (110 g). ±0.001 g (500 g). ±0.1 g (60 kg).[2]

OVERLOAD PROTECTION None required. Over loading will result in no more than the scale pan remaining at the extreme of its travel.

SUPPLY VOLTAGE Either self-contained battery operated or supplied from normal mains.

SENSITIVITY 0.001 mg (4 g), 0.01 g (5.5 kg) and 1 g (60 kg).[3]

CROSS SENSITIVITY Although great care is taken to ensure an insignificant error when loaded eccentrically it is good practice to load centrally.

TEMPERATURE RANGE + 10 to + 30°C..

TEMPERATURE SENSITIVITY $±2 \times 10^{-6}$ g/°C, subject to capacity.

SIGNAL CONDITIONING Self-contained with a direct reading display and features such as tare subtraction and data storage and remote outputs.

BENEFITS AND LIMITATIONS Fine examples of electrical devices which have the capability close to their precision mechanical counterparts but with numerous other benefits.

COMMENTS [1]An integral calibration mass is applied to allow setting up the servo balance and for performance checking.
[2]The values of error are usually expressed in grams hence they are dependent upon the selected range. Therefore these are but examples. All values in parenthesis identify the relevant range.
[3]Dictated by the display capability of the balance. The values are but examples. All values in parenthesis identify the relevant range.

SUPPLIERS EUR2; GEC; MAR.

Mass, weighing

WE2. Balance, weight comparator

OPERATING PRINCIPLE Essentially a servo device comprising an electrically energised inductive coil located around a magnetic load-carrying core.

When load is applied, causing movement of the core, a detector senses this positional change and induces a change in the coil current sufficient to restore the core to its original position.[1]

Weighing is achieved by sequentially comparing a known reference weight and the sample.

The visual display presents the difference between the two.

APPLICATIONS These devices are weight comparators with a very high order of accuracy.

Thus they may be used to provide the calibration and certification of masses.

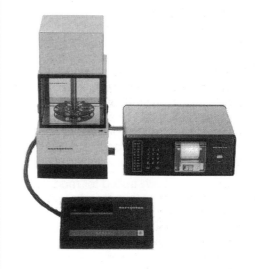

Courtesy Sartorius AG

RANGE LIMITS 0 – 5 g to 0 – 60 kg in a number of ranges.[2]

ERROR Repeatability: ±1 μg (5 g range); ±15 μg (200 g range); ±50 mg (60 kg range).

OVERLOAD PROTECTION None required. Over loading will result in no more than the scale pan remaining at the extreme of its travel.

SUPPLY VOLTAGE Normal mains voltage.

SENSITIVITY ±0.1 μg (5 g range); ±1 μg. (200 g range); ±10 mg (60 kg range).

CROSS SENSITIVITY Insignificant.

TEMPERATURE RANGE + 10°C to + 30°C.

TEMPERATURE SENSITIVITY Errors are accommodated within the overall error above.

SIGNAL CONDITIONING This is incorporated within the balance and includes the visual display unit.

BENEFITS AND LIMITATIONS Of a very high order of accuracy. Being a comparator these are true mass balances.

COMMENTS [1]The position detector may be optical, inductive or capacitive.
[2]In order to achieve the maximum precision of measurement the weighing process as such is not necessarily continuous but structured to measure about selected values or weighing ranges. For example in the 200 g total range mentioned above, the weighing ranges could be at 10, 20, 50, 100 & 200 g with the range window being ±0.5 g..

SUPPLIERS EUR2; SAR1; SAR2.

Mass, weighing

WE3. Load cell, industrial platform weighing

OPERATING PRINCIPLE Comprises a single low profile load cell used to support a rugged platform. Bending, shear and compression load cells may be used dependent upon range.   These devices may be floor or bench mounted with the former being more rugged with the electronic display either remote from the platform such as pillar mounted to one edge of the platform or integral with the balance, usually immediately below the scale pan.

APPLICATIONS Ideally suited for modern production and packaging where the accuracy requirements are less severe than that of precision balances.

Courtesy Marsden Weighing Machine Group Ltd

RANGE LIMITS Scale pan: 0 – 500 g to 0 – 3000 kg. Specialised weighing: up to 50,000 kg.[1,2]

OVERLOAD PROTECTION At least ×1.25 of full range with the possibility of overload stops fitted for the lower ranges. The visual display unit may identify overload conditions.

ERROR Nonlinearity, hysteresis, repeatability and temperature coefficient combined: ±0.02% to 0.033% of full range up to 500 kg. ±0.03% to 0.1% above 500 kg.

SUPPLY VOLTAGE 115 to 240 V ac mains supply or battery operated.

SENSITIVITY Resolution: between 0.02% and 0.33% of full range.

CROSS SENSITIVITY Insignificant, although it is good practice to ensure that loads are placed central on the scalepan or platform.

TEMPERATURE RANGE Compensated: –10°C to +70°C. Operating: –10°C to +80°C.

TEMPERATURE SENSITIVITY Zero: ±0.005% of full range/°C. Range: ±0.005% of full range/°C.

SIGNAL CONDITIONING Supplied with a wide range of options including visual display, printers, additional information storage, and electrical signal output.

BENEFITS AND LIMITATIONS These are amongst the best of modern loadcell technology combined with electronic capability.

COMMENTS Tare weight subtraction may be available also the option of establishing the calibration against a known mass or a representative group of items such as a set of nuts and bolts.
[1]Such as hopper weighing.
[2]Although operating down to "zero" there will be some lower value for which Weights and Measures approval will be required. These lower values are often very low such as ±1% of full range and certainly generally better than ±5% of full range.

SUPPLIERS ACT; EUR2; GEC; MAR; MEC; PRE; TEC; TED.

Mass, force

The process of weighing is one of the earliest forms of measurement (see Mass, weighing). Today by far the most common method of measuring force or load is by some form of metallic spring element. This required the development of metals which could conform to the linearity predicted by Hooke's Law to a high degree of precision. The demand for these devices is continuous, great and increasing.

Without doubt the vast majority of force measuring devices use **resistance strain gauges** as the sensing element (see Strain). They may be categorised by the mode in which the strain or deformation in the spring element is initiated; by bending, shear or direct (tensile or compressive) loading. Usage may be confined to a single direction of loading or in both directions along the chosen axis (unidirectional or bidirectional). Loading is usually identified as tensile or compressive which is related to the mode of use rather than the manner in which the deformation is produced.

The choice is mainly influenced by the magnitude of the load or force with **bending** being chosen for the smaller values and **direct loading** for the largest with **shear** devices being used in the middle ranges. There is considerable overlap for each type category. The very nature of the deformation results in an inherent nonlinearity for all but the shear devices. However such error is usually small in a well-designed device.

The choice of a particular type of device should be made with care especially when used in a complex force field. This arises because the cross sensitivity of the load cell can be high with the serious prospect of going undetected. At least one laboratory will undertake tests to establish the performance of a load cell when subject to a bending moment about the line of axis of the force using their deadweight machines.

Measurement of smaller force ranges require special techniques of which both the **linear variable differential transformer (LVDT)** and the **unbonded strain gauge** are readily available devices.

The **multicomponent force balance,** especially for three or more forces and/or moments, is a specialised requirement and is often purpose-built. The examples overleaf are but a selection available on a commercial basis.

In building a **multicomponent strain gauge balance,** a number of elements are each configured to measure a particular single axis of force or moment whilst being generally insensitive to all of the others. This may be achieved either by ensuring that each particular element receives insignificant deformation or strain from these unwanted components or that the sensing element is unresponsive to such deformation. It is because the **bonded resistance strain gauge** has a capability to deal with the latter that the large majority of multicomponent force balances use them.

Ideally each element should be stiff in one axis, the measuring one, and at least one axis should be both weak and insensitive to loading. Often this is difficult to achieve especially where some other constraint such as size is imposed. Because of such limitations, some balances do experience significant interactive effects which need to be determined by a complex calibration technique.

With a system of multiple forces and moments it is common practice to machine the whole transducer element from one piece of metal, since it is found that bolted or welded assemblies result in unwanted and erratic movements of the various parts giving rise to undesirable errors. Indeed the only type of bolt-together balance which will perform with acceptable characteristics uses shear plates where small slippage between the elements may result in small changes of output of individual elements but not the overall measurement result. (However it remains good practice to ensure that the chosen attachment method of the shear plate reduces these effects to a minimum.)

The use of **piezo-electric crystals** also provides a method of measuring both single and multiple forces in a multicomponent mode. They have an important role to play in the measurement of dynamic and quasi-static forces since they have a high natural frequency and, inherently, can carry high static loads without it affecting the dynamic characteristics.

The quartz crystal elements may be configured to respond to both direct loading and to shear. Hence flat disc devices may be so stacked that at one location all three mutually perpendicular forces may be measured. By using a multiplicity of locations any number of forces and moments can be determined. Hence the design of such multicomponent force balances is entirely different to that of the other type discussed above.

For single component devices, calibration is generally achieved by deadweight loading, universal testing machines or by a combination of deadweight and hydraulic amplification.

Calibration of multicomponent devices are much more sophisticated and time consuming and it may be necessary to construct a calibration rig which provides multiforce/moment capability for a single balance.

The degree of detailed calibration is dependent upon the required accuracy and on the degree of interaction. The lower the latter, the less calibration is required to establish a given performance.

For piezo-electric devices, the applied loads must be dynamic which poses even greater problems although beneficially the elements have an inherently low interaction.

Wherever possible multicomponent balances should be purchased complete with a calibration history and then it is necessary only to monitor the stability of the various elements.

Making a choice

The detailed selection of a load cell is one usually controlled by the force application, the desired accuracy and the size.

From the Contents listed below it will be seen that there are clearly defined bands of measurement for each type of transducer but there is a considerable overlap between most types. It is these overlaps which emphasise the necessity of making a thorough assessment. Where the application is one where only dynamic or quasi-static measurement is required then piezo-electric devices are most likely to be the first choice. For the remainder, range, cost, and the operating environment may well make the decision self-selecting.

NAMAS Accredited Laboratories: 0003, 0019, 0052, 0054, 0066, 0086, 0090, 0101, 0106, 0127, 0132, 0135, 0142, 0148, 0154, 0157, 0159, 0167, 0180, 0188, 0189, 0207, 0231, 0232, 0251, 0252.

The vast majority of these Laboratories offer a Testing Machine calibration service.

Contents: Mass, force

FO1. Load cell, bonded resistance strain gauge (bending)
5 N to 50 kN (tension and compression).

FO2. Load cell, bonded resistance strain gauge (direct loading)
50 kN to 5000 kN (tension and compression).

FO3. Load cell, bonded resistance strain gauge (shear)
2.5 kN to 200 kN (tension and compression).

FO4. Load cell, linear variable differential transformer (LVDT)
0.1 N to 0 – 1000 N (tension and compression).

FO5. Load cell, magnetostrictive
4 kN to 60 MN (compression only).
60 kN to 1.6 MN (tension only).

FO6. Load cell, piezo-electric
0.5 kN to 120 kN (tension and compression).
20 kN to 1000 kN (compression only).

FO7. Load cell, piezo-resistive
0.1 N to 10 kN (tension and compression).

FO8. Load cell, unbonded strain gauge
0.6 N (max) (tension and compression).

FO9. Load cell, vibrating wire
100 kN to 2 MN (compression only).
3 MN to 6 MN (compression only).

FO10. Load shackle, bonded resistance strain gauge
50 kN to 6000 kN (tension only).

FO11. Multicomponent force balance, bonded resistance strain gauge
Various ranges.

FO12. Multicomponent force balance, piezo-electric
Various ranges.

FO13. Wire rope tension
10 kN to 0 200 kN (tension only).

Mass, force

FO1. Load cell, bonded resistance strain gauge (bending)

OPERATING PRINCIPLE These devices are so configured that part or parts of a metal element experience some form of bending.
At these locations resistance strain gauges are bonded so as to respond to the strain and hence the load.
There are various chosen shapes and include cantilever, S-shape and box or ring sections.
The chosen material is usually either aluminium or steel but other materials are used occasionally.

APPLICATIONS Attached by screw thread or by simple location and used over a very wide range of applications from weighing to calibration.

Courtesy Thames Side Scientific Co Ltd

RANGE LIMITS 0 – 5 N to 0 – 50 kN (tension and compression).

ERROR Nonlinearity: ±0.03% of full range. Hysteresis: ±0.02% of full range. Repeatability: ±0.02% of full range.[1]

OVER-RANGE PROTECTION ×2 of full range without loss of calibration and ×3 of full range without structural failure is common.

SUPPLY VOLTAGE 10 V dc or ac (rms) maximum is normal.

SENSITIVITY Between 2 to 3 mV for full range/V of input supply.

CROSS SENSITIVITY Seldom quoted, but transverse sensitivity may be sufficient to affect the claimed accuracy unless mechanically protected from other forces or moments.

TEMPERATURE RANGE Compensated: –20°C to +80°C is most common.
Operating: –40°C to +100°C.

TEMPERATURE SENSITIVITY Zero: ±0.003% of full range/°C. Range: ±0.0015% of full range/°C.

SIGNAL CONDITIONING Operate into strain gauge amplifiers or digital voltmeters, augmented with a stable power supply (ac or dc).

BENEFITS AND LIMITATIONS Some devices may be used in both tension and compression loading but others are specifically designed for unidirectional loading alone.

COMMENTS With a considerable number of manufacturers and a wide range of shapes, there will almost certainly be a device sufficient for most needs.
[1] The Error values are typical but do depend upon the mode of bending chosen. Individual values of the three errors is not consistent between manufacturers. However the combined value is generally below ±0.1% of full range. Some manufacturers will select transducers from their production, improving the error by up to a factor of ×10.

SUPPLIERS ACT; ENT; HYD; MAY; NEG; NOB; RDP; SCA1; SCA2; TEC; TED; THA.

Mass, force

FO2. Load cell, bonded resistance strain gauge (direct loading)

OPERATING PRINCIPLE A column, or columns, has bonded to it (them) a number of resistance strain gauges which respond to the direct strain imposed by the loading.
For tension, measurement has one column but for compression the load may be shared by a number of them. Steel is the preferred material. In compression, the load is introduced to the load cell through some form of centrally located pivot or ball seating, providing loading consistency.[1]

APPLICATIONS Simply attached and located and used over a very wide range of applications including production weighing and calibration.

Courtesy Negretti Automation Ltd

RANGE LIMITS From 0 – 50 kN to 0 – 10,000 kN or more (tension and compression).[2]

ERROR Nonlinearity: ±0.03% of full range. Hysteresis: ±0.02% of full range. Repeatability: ±0.015% of full range.[3]

OVER-RANGE PROTECTION Generally at least ×1.5 of full range without loss of calibration should be expected.

SUPPLY VOLTAGE 10 V dc or ac (rms) maximum is normal.

SENSITIVITY Between 1 and 3 mV for full range/V of input supply.

CROSS SENSITIVITY Seldom quoted, but transverse sensitivity may affect the claimed accuracy unless mechanically protected from other forces or moments, particularly in compression.

TEMPERATURE RANGE Compensated: –20°C to +80°C are common.
Operating: –40°C to +100°C.

TEMPERATURE SENSITIVITY Zero: ±0.003% of full range/°C.
Range: ±0.0015% of full range/°C.

SIGNAL CONDITIONING Operate into strain gauge amplifiers or digital voltmeters, augmented with a stable power supply (ac or dc).

BENEFITS AND LIMITATIONS Usually available for loading in either compression or tension only although some are bidirectional.

COMMENTS [1]It must not be assumed that the swivelling action continues once the cell is under load, since friction will usually inhibit this. [2]Note that the higher values are limited to compression alone. [3]Better performance may be achieved but this requires that a high precision large capacity calibration facility such as a dead weight machine is available; this probably being the ultimate limitation in providing high accuracy.

SUPPLIERS ACT; AJB; EUR3; MAY; NAT; NEG; PIA; RDP; SCA1; SCA2; SEN2; STR; TEC; TED; THA.

Mass, force

FO3. Load cell, bonded resistance strain gauge (shear)

OPERATING PRINCIPLE These devices, usually of steel or aluminium construction, contain an element or elements such that under loading they experience a shear strain.
Resistance strain gauges are bonded to these shear elements, the gauges responding to the principal strains set up in the shear strain field.
This requires that the gauges be placed at an angle of 45° to the shear strain.

APPLICATIONS Attached by screw thread or by simple location and used over a very wide range of applications including weighing and calibration. Particularly suitable in a multi-axis force field.

Courtesy AJB Associates Ltd

RANGE LIMITS 0 – 2.5 kN to 0 – 200 kN are common (tension and compression).

ERROR Nonlinearity: ±0.03% of full range. Hysteresis: ±0.02% of full range. Repeatability: ±0.01% of full range.[1]

OVER-RANGE PROTECTION ×2 of full range without loss of calibration and ×3 of full range without structural failure is common.

SUPPLY VOLTAGE 10 V dc or ac (rms) maximum is normal.

SENSITIVITY Between 1 and 2 mV for full range/V of input supply.

CROSS SENSITIVITY Transverse sensitivity is usually very good. with claims typically between "insignificant" to 1 in 250 side force rejection.

TEMPERATURE RANGE Compensated: –20°C to +80°C are common. Operating: –40°C to +100°C.

TEMPERATURE SENSITIVITY Zero: ±0.003% of full range/°C. Range: ±0.005% of full range/°C.

SIGNAL CONDITIONING Operate into strain gauge amplifiers or digital voltmeters, augmented with a stable power supply (ac or dc).

BENEFITS AND LIMITATIONS The advantage of "shear" is the ease with which a design can be made relatively immune to other load directions.

COMMENTS [1]The Error values are typical. However individual values of the three errors is not consistent between manufacturers but the combined value is generally below ±0.1% of full range. Some manufacturers will select transducers from their production, improving the error by up to a factor of ×10.

SUPPLIERS ACT; AJB; EVR3; MAY; NEG; RDP; SCA1; SCA2; TEC; THA.

Mass, force

FO4. Load cell, linear variable differential transformer (LVDT)

<table>
<tr>
<td>

OPERATING PRINCIPLE An LVDT consists of an energised central primary winding flanked by two secondary windings.

These three coils are in effect two electrical transformers and their coupling can be varied by a central core being moved through the windings. The difference in output between the two secondary windings provides a direct linear measurement of the core displacement

By being mounted into a flexible element, as load is applied the deflection is detected.

</td>
<td rowspan="2">

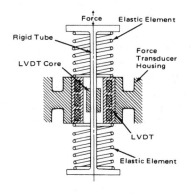

Schematic diagram of LVDT used in a force transducer

Courtesy Lucas Schaevitz Ltd

</td>
</tr>
<tr>
<td>

APPLICATIONS Wherever a high stability low force measurement is required and/or where the ac energisation is convenient such as for aircraft use.

</td>
</tr>
</table>

RANGE LIMITS 0 – 0.1 N to 0 – 1000 N (tension or compression).

OVER-RANGE PROTECTION Can be fitted with overload stops to provide a high overload capability.

ERROR Nonlinearity: ±0.2% of full range. Repeatability: ±0.1% of full range.

SUPPLY VOLTAGE Voltage: 1 to 5 V ac to the LVDT
Frequency: between 50 Hz and 25 kHz.[1]

SENSITIVITY Between 80 mV rms/V rms to 250 mV rms/V rms input or ±5 V dc (with signal conditioning).

CROSS SENSITIVITY The general design ensures an extremely low transverse sensitivity.

TEMPERATURE RANGE Operating: –50°C to +90°C. Storage: –50°C to +150°C.

TEMPERATURE SENSITIVITY ±0.05% of full range/°C.

SIGNAL CONDITIONING 15 V dc. (Usually supplied as part of a complete transducer system.)

BENEFITS AND LIMITATIONS They can incorporate mechanical adjustment of the zero output position which allows the accommodation of reasonably large tare loads.

COMMENTS LVDTs allows conversion of existing mechanical systems, such as proving rings, into electrical ones.
[1]Preferred choices are usually within the 400 Hz to 10 kHz band, but is related to the use. The higher frequencies give higher dynamic response but may result in electromagnetic radiation problems. Or the choice may depend upon one of convenience. For example 400 Hz is standard supply on modern aircraft.

SUPPLIERS LUC1; LUC2.

FO5. Load cell, magnetostrictive

OPERATING PRINCIPLE Lines of force of a magnet are distorted when mechanically stressed. If a stack of lamina metal plates are pierced by four holes, two electrical windings may be constructed in a cruciform pattern, using opposing hole pairs to form a transformer of negligible coupling. When the laminations are stressed by a force at 45° to the windings axes, the magnetic distortion causes a signal to appear at the secondary proportional to the force.[1]

APPLICATIONS Extremely robust, answering many applications for weighing on heavy machinery such as cranes, offshore rigs, steel mills or shipyards.

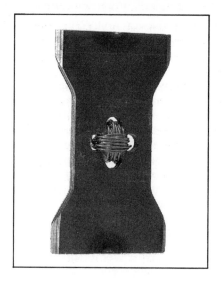

Photograph of transducer load cell showing cross-windings mounted on several plates welded together

Courtesy ABB Automation

RANGE LIMITS Compression: 0 – 4 kN to 0 – 60 MN. Tension: 0 – 60 kN to 0 – 1.6 MN.

ERROR (Tension and compression) nonlinearity: ±0.5% of full range. Hysteresis: ±0.2% of full range. Repeatability: ±0.05% of full range.

OVER-RANGE PROTECTION At least ×3 without change of calibration and ×5 without mechanical damage.

SUPPLY VOLTAGE Usually provided complete with signal conditioning and operated from normal mains supply.

SENSITIVITY Typically 5 V dc for full range output after signal conditioning.

CROSS SENSITIVITY Low. Indeed one feature claimed for these devices is their insignificant transverse sensitivity.

TEMPERATURE RANGE Compensated: +20°C to +90°C. Operating: –40°C to +90°C. (Up to 110°C for short periods).

TEMPERATURE SENSITIVITY Zero: ±100 ppm/°C. Range: 100 ppm/°C.

SIGNAL CONDITIONING Purpose-built to provide the transformer energisation and signal conditioning.

BENEFITS AND LIMITATIONS It is possible to use existing structural elements as the spring element, making them particularly suitable for retrofitting

COMMENTS [1]These devices come in a variety of shapes including annular, circular and square. They may be constructed for special applications such as monitoring forces on pedestal bearings used in rolling mills.

SUPPLIERS ABB2; ABB3; INT.

Mass, force

FO6. Load cell, piezo-electric

OPERATING PRINCIPLE A piezoelectric crystal exhibits the property that when deformed it yields an electrical charge, measured in Coulombs, proportional to that deformation. Thus by arranging that the load is transferred through such a crystal a load cell may be constructed.[1]
If the deformation is maintained, the charge will gradually decline due to parasitic high resistance paths. Thus these devices are not suitable for static measurement.

APPLICATIONS Incorporating the required attachments, they are used wherever dynamic force characteristics are required.
A wide range of uses from orthopaedic to machinery vibration.

Courtesy Kistler Instruments Ltd

RANGE LIMITS 0 – 0.5 kN to 0 – 120 kN, tension and compression.[2] 0 – 20 kN to 0 – 1000 kN compression only (load washers).
ERROR Nonlinearity: ±0.5% of full or partial range. Hysteresis: ±0.5% of full or partial range.
OVER-RANGE PROTECTION ×1.1 to ×1.2 of full range.[2]
SUPPLY VOLTAGE Self-generating.
SENSITIVITY 2 to 4 pC/N.
CROSS SENSITIVITY Transverse sensitivity is dependent upon design, installation and use. Values can be up to ±1% of the operating range.
TEMPERATURE RANGE Operating: –196°C to + 120°C.
TEMPERATURE SENSITIVITY 0.01% to 0.02% of full range/°C.
SIGNAL CONDITIONING Charge amplifiers are required to prevent signal deterioration. For partial range use, good conditioning will be essential.
BENEFITS AND LIMITATIONS They are capable of measuring small dynamic loads whilst subjected to high static forces. High stiffness ensures a high natural frequency.
COMMENTS [1]Piezo-electric load cells are available in both tension and compression and some are in the form of load washers which allow for ready installation into a bolted structure. [2]Piezo-electric devices may be used over a part of a selected range without loss of performance and manufacturers supply calibrations for these partial ranges also resulting in extremely high overload capability.
SUPPLIERS KIS1; KIS2; KIS3; KIS4.

FO7. Load cell, piezo-resistive

OPERATING PRINCIPLE Piezo-resistive devices use semiconductor strain gauges.
These are elements of silicon crystal and have a gauge factor which can be well in excess of 100.
Semiconductor gauges are bonded to the body of the transducer much as for conventional bonded resistance strain gauges with transducer shapes conforming generally to those for conventional strain gauge transducers. See also "Strain".
The chosen material is usually steel or aluminium.

APPLICATIONS The small size and high natural frequency ensures a wide range of uses from control loop systems to dental research.

Courtesy Entran Ltd

RANGE LIMITS $0 - 0.1$ N to $0 - 10$ kN, tension and compression.	
ERROR Nonlinearity: ±0.1% of full range. Hysteresis: ±0.05% of full range.	
OVER-RANGE PROTECTION From ×1.25 to ×4 of full range.	
SUPPLY VOLTAGE Gauge bridge: 5 V to 15 V dc.[1]	
SENSITIVITY Bridge output: 50 mV to 250 mV for full range output.[2]	
CROSS SENSITIVITY Not often quoted and influenced by design shape.	
TEMPERATURE RANGE Compensated: 0 to +55°C.[3] Operating: –40°C to +160°C.	
TEMPERATURE SENSITIVITY Zero: ±0.03% of full range/°C. Range: ±0.03% of full range/°C.	
SIGNAL CONDITIONING Strain gauge amplifiers or digital voltmeters are suitable but it is important to operate at the manufacturer's preferred bridge supply voltage.	
BENEFITS AND LIMITATIONS These devices add to the capability of small size and small force measurement. They have a high natural frequency whilst still maintaining a high voltage output.	
COMMENTS [1]Some devices come complete with signal conditioning. Under these circumstances a dc supply voltage such as 5 V dc or ±15 V dc is required. [2]5 V dc for full range is common when fitted with integral signal conditioning. [3]Semiconductor devices are usually compensated over 50°C bands because of the non-linear temperature response. The compensation will be undertaken for a set input voltage. Changes to this voltage may affect the quality of the temperature compensation.	
SUPPLIERS ENT; KUL1; KUL2; MAY.	

FO8. Load cell, unbonded strain gauge

OPERATING PRINCIPLE A metal filament, formed into a grid or grids, is supported at each end by insulating carriers.

When loaded the resultant strain within the wire results in a resistance change proportional to the applied force.

By prestraining, the filaments can either reduce or increase their strain, thus allowing the formation of a Wheatstone bridge. The whole is carried by spring elements, such as diaphragms, one at each end.

APPLICATIONS Specialised use where very small force measurement is required.

They may be adapted for other measurements such as low pressure.

Courtesy Scaime Techno Talent

RANGE LIMITS 0 to 0.6 N, tension and compression.

ERROR Nonlinearity and hysteresis combined: ±0.15% of full range.

OVER-RANGE PROTECTION If fitted with overload stops, the protection will be high.

SUPPLY VOLTAGE Typically 5 volts ac or dc.[1]

SENSITIVITY 12 mV of full range/V of input supply.

CROSS SENSITIVITY The grid design should ensure that transverse sensitivity is insignificant or at least extremely low.

TEMPERATURE RANGE Compensated: –15°C to +65°C. Operating: –25°C to +90°C. Storage: –50°C to +95°C.

TEMPERATURE SENSITIVITY Zero: ±0.01% of full range. Range: ±0.01% of full range.

SIGNAL CONDITIONING As strain gauge devices, albeit unbonded, they can operate into strain gauge amplifiers or digital voltmeters.

BENEFITS AND LIMITATIONS They enjoy all those benefits which can be attributed to resistive transducers in general except possibly for the lower supply voltage.

COMMENTS The unbonded strain gauge enjoyed great popularity in the past before the bonded resistance gauge had reached its present excellence.
[1]This type of resistance element receive no cooling from a parent body. Thus it is particularly important to follow the manufacturers recommendations for the supply voltage.

SUPPLIERS SCA1; SCA2.

FO9. Load cell, vibrating wire

OPERATING PRINCIPLE When plucked a pretensioned wire will vibrate at its natural frequency.[1]
If the wire is then subject to a change of tension the frequency will change. The wire is "plucked" by a pulse of dc voltage applied to an electromagnet, centrally located, and the resulting vibratory signal picked up by the same or a second electromagnet. By using this device to measure deformation of a parent metal element, load is measured.

APPLICATIONS This type of load cell is most commonly used in civil engineering applications such as loads on piles or tension in cable anchors or rock bolts.

Courtesy Soil Instruments Ltd

RANGE LIMITS 0 – 100 kN to 0 – 2 MN, compression only. 0 – 3 MN to 0 – 6 MN, compression only.[2] 0 – 100 kN to 0 – 2 MN, tension only.[3]

ERROR Nonlinearity and hysteresis: ±0.5% of full range.

OVER-RANGE PROTECTION Yield is ×2.5 of the working range.

SUPPLY VOLTAGE Self-generating. 10 to 50 V dc pulse voltage.

SENSITIVITY Dependent upon the ability to measure changes of frequency.

CROSS SENSITIVITY With proper installation, transverse sensitivity should be low.

TEMPERATURE RANGE Operating: typical natural environmental conditions.

TEMPERATURE SENSITIVITY Low, with the wire having the same coefficient of expansion as the body of the load cell.[4]

SIGNAL CONDITIONING Purpose-built; the display provides a linear correlation between load and output. This same unit will provide the "plucking" voltage.

BENEFITS AND LIMITATIONS For long-term use, it is both robust and highly stable but suitable only for static or quasi-static measurement.

COMMENTS [1] The relationship is: Frequency = $(t.g)^{0.5}/2L_w\sigma$ where t = tension, g = gravity, L_w = wire length and σ = density of wire.
[2] Three transducers combined in parallel to form a single load cell.
[3] Three load cells and of tubular form allowing the fastening of a rod or cable concentrically through the centre enabling tensile measurement such as cable loads.
[4] It is possible to include a thermometer to correct residual error.

SUPPLIERS ROC; SOI.

Mass, force

FO10. Load shackle, bonded resistance strain gauge

OPERATING PRINCIPLE A conventional wire shackle has resistance strain gauges bonded to its shackle pin. These gauges are located and oriented to respond to the shear loading immediately on either side of the load application point, with the gauged positions protected from the risk of damage from the shackle and water ingress.
Commercial steel shackle may be converted subject to the pin being located positively in a fixed angular position.

APPLICATIONS Wide ranging such as on crane hooks, tug tow-line forces, ship moorings and offshore installations.

Courtesy European Monitoring Services

RANGE LIMITS 0 – 50 kN to 0 – 6000 kN, tension only.

ERROR Nonlinearity and hysteresis: ±1.0% of full range.

OVER-RANGE PROTECTION ×1.5 overload without transducer damage and ×4 of shackle safe working load is typical.[1]

SUPPLY VOLTAGE 10 V to 15 V.

SENSITIVITY 1.5 to 2.0 mV for full range/V of input supply.

CROSS SENSITIVITY Should not respond to other forces but care must be taken that the gauges are correctly positioned.[2]

TEMPERATURE RANGE Operating: –10°C to +50°C.

TEMPERATURE SENSITIVITY ±1% of full range over normal environmental temperatures.

SIGNAL CONDITIONING Systems may be supplied complete with a purpose-built signal conditioning system; either hard wired or using telemetry.

BENEFITS AND LIMITATIONS The use of shackles provides a device which requires no special fittings to introduce it into any cabling system.

COMMENTS [1]If metal is removed from the pin to allow the proper installation of the gauges, the safe working load will not be the same as that normally used for an equivalent unmodified shackle.
[2]These shackles are normally used as part of a wire or cable connection. Hence they measure the tension of that wire. This may not be the desired force requirement without correction being made for the wire angle.

SUPPLIERS ACT; EUR2; STR.

FO11. Multicomponent force balance, bonded resistance strain gauge

OPERATING PRINCIPLE The flexural element(s) – usually in bending or shear, but occasionally in direct mode – are designed to give a maximum response to a single moment or force whilst giving the minimum response to all others.

Although a balance can measure between two and six components the natural division is either three or six components.[1]

APPLICATIONS Designed generally for specific tasks, usually for research and including hydrodynamics, aeronautics, vehicles and orthopaedic applications including human prosthesis.

Courtesy Hydraulics Research Ltd

RANGE LIMITS Wide ranging but generally low forces and moments.

ERROR Subject to calibration accuracy and degree of interaction.

OVER-RANGE PROTECTION Not normally designed with excessive overload protection capability.[2]

SUPPLY VOLTAGE 5 to 10 V dc or ac rms..

SENSITIVITY Outputs are generally lower than conventional load cells.

CROSS SENSITIVITY Dependent upon design.[3]

TEMPERATURE RANGE Generally ambient conditions except for specialised cases.

TEMPERATURE SENSITIVITY Dependent upon gauge selection.

SIGNAL CONDITIONING The system will be multichannel and there may be more channels than components of force being measured.

BENEFITS AND LIMITATIONS The major problems are in calibration to establish the output and transverse sensitivity to high precision.

COMMENTS [1]In aeronautical terms these are drag, yaw moment and side force as one grouping and lift, pitch moment and roll moment as the other, although their spatial attitude is not confined to these accepted directions.
[2]It is essential to define precisely what is meant by the "quoted" ranges.
[3]The calibration of a balance to establish the transverse sensitivity is a time-consuming task.

SUPPLIERS ACT; CON; HYD.

Mass, force

FO12. Multicomponent force balance, piezo-electric

OPERATING PRINCIPLE A piezoelectric crystal exhibits the property that when deformed by a force it yields an electrical charge, measured in Coulombs, proportional to that force. Hence by using a number of crystals in specific spatial configurations, a multicomponent force balance is produced. If the forces are maintained the charges will gradually decline due to parasitic high resistance paths. Thus these devices are not suitable for static measurement.[1]

APPLICATIONS Admirably suited to vibratory or oscillating tasks such as metal machining and biomechanical applications with the benefit of carrying very high static forces and moments.

Courtesy Kistler Instruments Ltd

RANGE LIMITS Two-component: Fz, –5 to + 20 kN. Mz, –100 to + 100 N.m. Four-component: Fx and Fy, –5 to + 5 kN. Fz, –5 to + 20 kN. Mz, –100 to + 100 N.m.	
ERROR Nonlinearity: ±1% of full range. Hysteresis: ±1% of full range.	
OVER-RANGE PROTECTION ×1.2 to 1.5.	
SUPPLY VOLTAGE Self-generating.	
SENSITIVITY Force: 2 to 8 pC/N. Moment: 150 pC/N.m.	
CROSS SENSITIVITY ±1% to ±3% of full range.	
TEMPERATURE RANGE Operating: –196°C to + 150°C.	
TEMPERATURE SENSITIVITY 0.02% of full range/°C.	
SIGNAL CONDITIONING These devices must be inputted to charge amplifiers. Special amplifiers are available which convert the charges into voltages and provide the desired forces and moments in the appropriate engineering units.	
BENEFITS AND LIMITATIONS Although limited to dynamic measurement this can be advantageous since the readings are not swamped with large static force outputs.	
COMMENTS [1]They are usually of disc form and so produced to provide potentials within the thickness of the disc (compression), and across the diameter (shear), and it is these two components of output which are used in multicomponent balances. [2]Piezoelectric devices are built to preset configurations rather than tailored to a set of requirements and there are a good range of sizes and shapes of which these are but examples. Up to six components may be configured.	
SUPPLIERS KIS1; KIS2; KIS3; KIS4.	

FO13. Wire rope tension

OPERATING PRINCIPLE A tensioned wire is deformed from its straight condition by the application of three equi-spaced forces.

The outer two are imposed by pulleys attached to a carrier or body of the device and the central force, acting in the opposing direction, is applied by a clamp rigidly attached to the wire rope.

This force is transferred to the carrier through a load cell.

The measured force is directly related to the tension force.

APPLICATIONS A simple clamp-on system used for a wide variety of wire tension systems.

It allows simple retro-installation and may be removed when not in use.

Courtesy PIAB Ltd

RANGE LIMITS Tension force: 0 – 10 kN to 0 – 200 kN.[1]

ERROR Nonlinearity: ±4% of full range. Repeatability: ±1.5% of full range.[2]

OVER-RANGE PROTECTION ×2 of full range.

SUPPLY VOLTAGE 10 V to strain gauge bridge supply.

SENSITIVITY Load cell output: 1.6 mV/V of input supply.
Signal conditioning output: 4 – 20 mA current loop.

CROSS SENSITIVITY Transverse sensitivity is low. There will be hysteresis from the effects of internal friction within the individual strands of the wire.

TEMPERATURE RANGE Operating: –40°C to + 70°C.

TEMPERATURE SENSITIVITY Low compared to overall performance.

SIGNAL CONDITIONING Signal conditioning will be either self-contained, complete with visual display, or provide a current loop output.

BENEFITS AND LIMITATIONS Having the capability of being fitted to installed cables makes them ideal for regular monitoring of any number of cables.

COMMENTS [1]To suit wire diameter of between 5 mm to 44 mm in selected ranges. Any single device will generally accommodate a variation of ±2 mm of the nominal wire size.
[2]The load cells, commonly bonded resistance strain gauge types, are likely to have a performance generally as quoted for load cells. However the values here identify the performance as fitted to the wire rope.

SUPPLIERS PIA.

Pressure, fluids

Conventional hydraulic or pneumatic pressure measurement developed to meet the needs of the Industrial Revolution, first as a liquid level in a pipe or channel and then into what is now the ubiquitous **Bourdon tube gauge.** Even without statistics to hand, it is reasonable to assume that this type of transducer is still the most common means of measuring pressure. However, electrical output pressure transducers have been in existence for some considerable time and because of modern needs for remote operation are becoming more and more common in many branches of measurement.

As with many electrical transducers, the method of measurement uses all of the common sensing elements such as **strain gauges** and **piezo-electric** devices. But apart from adaptations or conversions of mechanical transducers, the usual practice is to use a **diaphragm** or **bellows** as the sensing element and determine its deformation as a measure of pressure.

Since much of the measurement of pressure is remote, housed in pipes or pressure vessels, the advantage of the electrical transducer to respond to rapid changes of pressure is of particular importance.

The **microphone** too is a type of pressure transducer. It became more than a mere scientific experiment with the evolution of the telephone and the network of wires which allowed a ready expansion of simple voice communication. Today the use of the microphone is commonplace and yet very few realise the intricacies and performance criteria of a modern microphone and its near relative the **hydrophone**. The essential similarity between the microphone and pressure transducers highlights the possible interaction of sound or noise upon pressure measurement.

Three types of pressure measurement are possible and are identified as follows:

   1. Gauge pressure This compares the applied pressure to that of the local environment; usually the barometric or atmospheric pressure.

   2. Absolute pressure This compares the applied pressure with essentially that of zero pressure using an internally sealed vacuum. (This is for example the requirement for measurement of barometric pressure using an aneroid barometer). There is a subset where this reference pressure may not be zero but some other preset value such as 1 atmosphere. These are called "sealed" devices.

   3. Differential pressure Here the output is the result of the difference of two applied pressures. Where pressure is applied to the differential side of the pressure device care must be exercised that the internal mechanism is able to accept the pressurising medium.

NOTE. Except where identified, all of the transducers listed overleaf will include devices suitable for both **liquid** and **gaseous** use. Since pressure acts normal to a surface there is no transverse sensitivity as such. Thus all of the comments under Cross Sensitivity highlight other potential problems.

For extremely high pressures sometimes a cylindrical pressure chamber is used whereby the extension of the walls of the cylinder is measured. In extreme pressure conditions the compressive deformation of a solid metal block has been used. However neither of these are considered here.

The problem of using a plain diaphragm is to ensure a rigid attachment at the perimeter, at least sufficient to maintain under repeated use the chosen form of deformation and hence provide adequate repeatability. This also controls the method of transducer mounting so that the rigidity is not further influenced by the choice of attachment clamp or flange.

An alternative to the diaphragm is the bellows device where a considerable displacement occurs when the pressure is applied. These are popular when extremely low pressure measurement is required but the penalty is that the natural frequency will be considerably lower. (Mechanical aneroid barometers use similar bellows where the motion is sufficient to move a pointer, or in the case of a barograph, a pen across a paper chart.) Bellows can be either of the "accordion" type or a corrugated flat disc.

Even within a particular style of pressure transducer there is an extremely wide range of shapes and sizes to suit each particular need such as oceanography, geology, hydraulic control systems or engine cylinder pressures. In keeping with the wide applications there is not only an enormous size range but a wide frequency response. Diameters as low as 1.5 mm and natural frequencies as high as 200 kHz are available when such attributes are required.

There is a corresponding large choice of pressure connections. Care must be exercised for liquid pressure measurement so that pockets of gas do not hinder the measurement particularly for the lower pressure ranges. Where possible it is best to mount these devices so that gases will "float" or be purged away naturally from the transducer's pressure surface.

In some applications the location of the transducer will affect the value recorded. This is because the transducer cannot distinguish between the static and dynamic pressure which may be present.

Since pressure is a combination of force and area, the calibration method will be limited by the ability to combine these two parameters. The most common method for measuring pressures from 150 kPa to 60,000 kPa is by Deadweight Testers. In essence these comprise a piston of known area operating in a liquid-filled cylinder. The outlet from this vertically- mounted piston/cylinder combination is connected to the pressure transducer and deadweights are applied to the free end of the cylinder.

Various techniques are used to improve the performance of these devices, in particular rotating the cylinder to reduce the effects of friction to a minimum. Orders of accuracy are ±0.04% of selected full range. (Note that because of self-weight there is a minimum pressure value which can be attained.)

Low pressure calibrations may be performed by comparison between the transducer and a liquid manometer, usually either water or mercury filled, when subject to a common pressure. Indeed, any device which has an appropriate calibration history may be used for comparison purposes.

In spite of international agreement on rationalising units to a common SI system it is still regular practice to measure in units related to the task. As an example, low pressures are quoted in "inches water gauge" because often the transducer is either measuring actual water height or is replacing a water-filled manometer. Here, however, we quote pressure in Pascals. 1 Pascal (Pa) = 1 N/m^2.

Pressure devices are commonly used for the measurement of liquid level or water depth. Indeed the quartz resonator described in this section is often used for precise measurement of water depth such as deep ocean tide heights and its quality has ensured its inclusion in this Section. However two liquid level pressure devices are also included under Length.

NAMAS Accredited Laboratories: 0013, 0060, 0072, 0099, 0123, 0127, 0128, 0138, 0139, 0166, 0173, 0176, 0202, 0208, 0221, 0222 and 0234.

Contents: Pressure, Fluids (including gases)

PR1. Bonded resistance strain gauge
10 kPa to 70 MPa.

PR2. Capacitance
125 Pa to 2500 Pa.

PR3. Capacitance, barometric
50 kPa to 106 kPa.

PR4. Linear variable differential transformer (LVDT)
62.5 Pa to 350 kPa.

PR5. Piezo-electric
20 MPa to 1000 MPa.

PR6. Piezo-resistive, semiconductor gauges
13 kPa to 170 MPa.

PR7. Piezo-resistive, silicon diaphragm
13 kPa to 200 MPa.

PR8. Quartz resonator
40 kPa to 69 MPa.

PR9. Remote capacitance transmitter
In excess of 35 kPa.

PR10. Resistance potentiometer
20 kPa to 300 kPa.

PR11. Thick-film resistance strain gauges
10 kPa to 20 000 kPa

PR12. Vibrating wire
0.7 MPa to 20 MPa.

Pressure, fluids

PR1. Bonded resistance strain gauge

OPERATING PRINCIPLE The pressure impinges upon a diaphragm which either has strain gauges bonded directly to its surface or a strain gauge sensing element coupled to it. The deformation of the diaphragm is detected by the strain gauges and hence is a measure of the applied pressure.
The diaphragm is usually of stainless steel.

APPLICATIONS Mounted by typical pipe connectors or sealed flanges, they have a wide use including the water industry, hydraulic and pneumatic monitoring and many research applications.

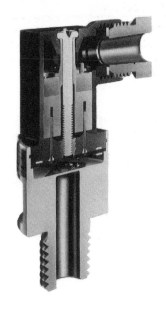

Courtesy UCC International Ltd

RANGE LIMITS 0 – 1,400 kPa to 0 – 35,000 kPa; 0 – 70 kPa to 0 – 70,000 kPa; 0 – 10 kPa to 0 – 35 kPa.[1]

ERROR Nonlinearity: ±0.15% of full range. Hysteresis: ±0.05% of full range.

OVER-RANGE PROTECTION Over-pressure: ×2 to ×5. Burst pressure: ×3 to ×20.[2]

SUPPLY VOLTAGE 10 V dc or ac (rms) maximum, unless part of the specification.

SENSITIVITY Between 2 mV/V and 3 mV/V of input supply.

CROSS SENSITIVITY Some devices may be sensitive to acceleration and vibration.

TEMPERATURE RANGE Compensated: –20°C to +90°C.[3]
Operating and storage: –55°C to +105°C.

TEMPERATURE SENSITIVITY Zero: 0.02% of full range. Range: 0.02% of full range

SIGNAL CONDITIONING Operate into strain gauge amplifiers or digital voltmeters; or signal conditioning equipment which is supplied complete with the transducer.

BENEFITS AND LIMITATIONS They cover a wide range of pressures and are configured for a comprehensive selection of uses.

COMMENTS [1]Selection is related to the application, such as full water immersion, or the type of measurement such as "gauge" or "absolute". These examples are typical range limits for a selection of designs and shapes.
[2]Burst pressure is the pressure leading to total failure resulting in leakage.
[3]Not necessarily applying to the connecting cables. For those transducers specially designed for water pressure measurement, often the operating temperature will be limited to between 0°C and 100°C.

SUPPLIERS ENT; LUC1; LUC2; MAY; PEN; RDP; UCC1; UCC2; UCC3; WES.

PR2. Capacitance

OPERATING PRINCIPLE A typical design comprises a capacitance constructed from a diaphragm within a sealed chamber.

The diaphragm is one plate of a differential arrangement with fixed capacitor plates being placed on either side, providing a half bridge circuit.

Pressure applied to either side of the diaphragm will result in a differential change of capacitance related to the pressure.

APPLICATIONS The electrical equivalent of the precision micromanometer and generally used for low air pressure measurements such as in wind tunnels or Pitot tubes.

Courtesy Airflow Developments Ltd

RANGE LIMITS	0 – 125 Pa to 0 – 2,500 Pa.
ERROR	Accuracy: ±1% of displayed reading.
OVER-RANGE PROTECTION	×10 of full range or higher
SUPPLY VOLTAGE	Battery powered including rechargeable versions.
SENSITIVITY	0 – 1 mA current output for full range.
CROSS SENSITIVITY	Insignificant for usual applications.
TEMPERATURE RANGE	Operating: 0°C to +40°C. Storage: –10°C to +60°C.
TEMPERATURE SENSITIVITY	±0.1% of reading/°C.
SIGNAL CONDITIONING	Supplied with its own signal conditioning, some of which is of great sophistication and is effectively a microprocessor.
BENEFITS AND LIMITATIONS	Convenient and simple to use.
COMMENTS	These devices can form part of a velocity or volume flow system when connected to the appropriate Pitot static device.
SUPPLIERS	AIR; RDP; ROS; TEC.

PR3. Capacitance, barometric

OPERATING PRINCIPLE Plates arranged in the form of a capacitance, located internally in an aneroid capsule, measure the deformation of the capsule as the atmospheric pressure changes.

To ensure the long-term stability required of barometers, a number of capsules may be used together to allow arithmetic manipulation of the electrical outputs, reducing the magnitude of random errors present in a single device.

APPLICATIONS Specifically designed for the measurement of barometric pressure and for use on airfields, ports and other meteorological tasks.

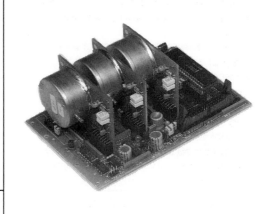

Courtesy Vaisala, Helsinki

RANGE LIMITS 50 kPa to 106 kPa.[1]

ERROR Accuracy: ±30 Pa over normal barometric range. Accuracy: ±50 Pa over full range.

OVER-RANGE PROTECTION None required for normal use.

SUPPLY VOLTAGE These devices are supplied with signal conditioning and display. All of the normal mains voltage supplies can be accommodated.

SENSITIVITY Display resolution: 10 Pa. Current output: 0 – 10 mA for full range output.

CROSS SENSITIVITY With temperature compensation and used in a proper location, cross sensitivity is insignificant.

TEMPERATURE RANGE Best operating: +5°C to +55°C. Storage: –40°C to +55°C.

TEMPERATURE SENSITIVITY Temperature compensation is incorporated and is contained within the values for Error above.

SIGNAL CONDITIONING A totally integrated system. It may provide additional capabilities such as a separate electrical output for additional data collection and storage.

BENEFITS AND LIMITATIONS Replacement for the mercury barometer, requiring no manual operation and providing automatic data logging.

COMMENTS [1]This is equivalent to 0.5 Bar to 1.05 Bar or 500 hPa to 1050 hPa; the latter being in hectoPascals where 1 hectoPascal = 1 mBar. (The hPa as a measurement is recommended for use by the World Meteorological Organisation.)

SUPPLIERS TEC; VAS.

PR4. Linear variable differential transformer (LVDT)

OPERATING PRINCIPLE An LVDT consists of an energised central primary winding flanked by two secondary windings.

These three coils are in effect two electrical transformers and their coupling can be varied by a central core being moved through the windings. The difference in output between the two secondary windings providing a direct linear measurement of the core displacement. By being coupled to a pressure capsule, as pressure is applied the deflection is detected.

APPLICATIONS Installed by flange mounting or screwed fittings, these devices are used for low pressure applications which require a high stability.

Courtresy Lucas Schaevitz Ltd

RANGE LIMITS 0 – 62.5 Pa to 0 – 350 kPa	
ERROR Nonlinearity, hysteresis and repeatability combined: ±0.5% to ± 0.75% of full range.	
OVER-RANGE PROTECTION ×1.5 of full range, low pressure. ×50 of full range, high pressure ranges.[1]	
SUPPLY VOLTAGE 10 to 32 V dc or a mains supply of 120/240 V ac (with integral signal conditioning).	
SENSITIVITY 5 to 10 V dc from signal conditioning output.	
TEMPERATURE RANGE Compensated: –5°C to +65°C. Operating: –40°C to +80°C.	
CROSS SENSITIVITY Because of the relatively high deformation they may be sensitive to vibration and acceleration.	
TEMPERATURE SENSITIVITY Zero: ±0.035% of full range/°C. Range: ±0.035% of full range/°C.	
SIGNAL CONDITIONING Typical of LVDT requirements but generally incorporated within the transducer housing.	
BENEFITS AND LIMITATIONS Most devices can be supplied in differential, gauge or absolute mode for at least the major part of the quoted ranges	
COMMENTS [1]This apparently disproportionate difference between the two quoted values is in no way a reflection on the quality of the design but more associated with the use. The very low pressure devices would normally be used in circumstances where overload is most unlikely whereas this may not be so for the higher pressure ranges.	
SUPPLIERS LUC1; LUC2; PLA; RDP.	

Pressure, fluids

PR5. Piezo-electric

OPERATING PRINCIPLE A piezo-electric crystal, when deformed, yields an electrical charge proportional to the deformation. For pressure measurement, a metal diaphragm is in direct contact with the fluid under pressure which deforms and transmits a load to a quartz crystal. If the deformation is maintained the resulting electrical charge will gradually decline due to parasitic high resistance paths. Thus these devices are not suitable for static measurements.

APPLICATIONS Used in a wide range of tasks, particularly hazardous ones, including engine combustion chambers, ballistics or for smaller fluctuations of pressure against a high static pressure background.

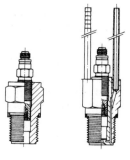

High pressure sensor with water-cooled adaptor

Courtesy Kistler Instruments Ltd

RANGE LIMITS 0 – 20 MPa to 0 – 200 MPa or 0 – 500 MPa to 0 – 1000 MPa[1]
ERROR Nonlinearity: ±0.5% to ±1.0% of full range.[2] Hysteresis: insignificant.
OVER-RANGE PROTECTION Typically ×1.2 to ×1.5 of full range.
SUPPLY VOLTAGE Self-generating.
SENSITIVITY –0.13 pC/kPa to –0.8 pC/kPa for full range output.
CROSS SENSITIVITY The inherent stiffness of the diaphragm/quartz assembly ensures a very low acceleration response.[3]
TEMPERATURE RANGE Operating: –196°C to +350°C.
TEMPERATURE SENSITIVITY +20°C to +100°C: ±0.5%; +20°C to +350°C: ±3%; +150°C to +250°C: ±1%. (All for full range).
SIGNAL CONDITIONING A suitable charge amplifier is required together with special cabling to inhibit signal leakage.
BENEFITS AND LIMITATIONS Care must be taken to understand the nature of these devices to be able to use them to their full potential.
COMMENTS Sensitivities between 0.00001 kPa/g and 0.0001 kPa/g are claimed for direct acceleration and significantly lower values for transverse acceleration. [1]Such as for ballistic measurements. [2]Because of the high linearity and low hysteresis, these devices may be supplied with partial calibrations at 1% and 10% of full range. It is claimed that the percentage errors are of the same magnitude for these partial ranges as they are for the full range. Partial ranges enhance the over-ranging capability considerably.
SUPPLIERS KIS1; KIS2; KIS3; KIS4; TEC.

Pressure, fluids

PR6. Piezo-resistive, semiconductor gauges

OPERATING PRINCIPLE Semiconductor gauges are bonded to a metallic element such as a diaphragm so that they respond to any deformation due to the applied pressure. See also Strain for a further description of this type of strain gauge.

APPLICATIONS With a high natural frequency, and high output they are well suited for many sophisticated static and dynamic pressure measurement applications.

Courtesy Entran Ltd

RANGE LIMITS 0 – 13 kPa to 0 – 170 MPa.

ERROR Nonlinearity and hysteresis: ±0.05% of full range. Repeatability: ±0.1% of full range.[1]

OVER-RANGE PROTECTION ×2 to ×10 dependent upon range.

SUPPLY VOLTAGE Up to 15 V dc and must be used at the manufacturer's stated values otherwise the temperature compensation may be rendered void.

SENSITIVITY 1 mV/V to 23 mV/V of supply.

CROSS SENSITIVITY Generally a very low acceleration response when compared to alternative choices.[2]

TEMPERATURE RANGE Compensated: –73°C to +230°C.[3] Operating: –196°C to +260°C.

TEMPERATURE SENSITIVITY Zero: ±0.013% to ±0.006% of full range/°C as the pressure range increases. Range: ±0.02% to ±0.04% of full range/°C.

SIGNAL CONDITIONING Dependent upon the desired operating frequency. Otherwise conventional strain gauge amplifiers may be employed.

BENEFITS AND LIMITATIONS A very wide choice of pressure range, size and capability in gauge, absolute and differential mode.

COMMENTS [1]Best values but generally overall accuracy will be between 0.5% and 1.25% of full range dependent upon type.
[2]Sensitivities in the pressure sensing direction as low as 0.00005% of full range/g for the higher ranges with cross axis sensitivity of 1/5 of this value are quoted.
[3]Normally they are conditioned to selected bands of approximately 50°C. This arises because of the temperature nonlinearity of semiconductor gauges.

SUPPLIERS ENT; EXP; MAY; MIC3; PRE; RDP; SAN.

Pressure, fluids.

PR7. Piezo-resistive, silicon diaphragm

OPERATING PRINCIPLE A silicon diaphragm has diffused into it inorganic atomically-bonded piezo-resistive strain gauges. The performance is similar to the semiconductor gauges but they are more difficult to protect from the pressure media and may shatter under particle impingement. Some types of transducer are available in a basic diaphragm form for installation into a suitable body prepared by the buyer.

APPLICATIONS With a high operating frequency up to 80 kHz, high output and, in some cases, extremely small size, they are used for many sophisticated static and dynamic pressure measurements.

Courtesy RS Components Ltd

RANGE LIMITS 0 – 13 kPa to 0 – 200 MPa.

OVER-RANGE PROTECTION ×2 over-pressure. ×3 burst pressure.

ERROR Nonlinearity and hysteresis: ±0.25% of full range. Repeatability: ±0.1% of full range.

SUPPLY VOLTAGE Up to 20 V dc. Must be used at the manufacturer's stated values otherwise the temperature compensation may be rendered void.

SENSITIVITY Between 7.5 mV/V and 10 mV/V of input supply.

CROSS SENSITIVITY Because of the inherent stiffness of the silicon diaphragm these devices have a very low acceleration response.[1]

TEMPERATURE RANGE Compensated: –55°C to +175°C.[2] Operating: –55°C to +175°C.

TEMPERATURE SENSITIVITY Zero: ±0.009% to ±0.036% of full range/°C. Range: ±0.02% to ±0.04% of full range/°C.

SIGNAL CONDITIONING Dependent upon the desired operating frequency. Otherwise conventional strain gauge amplifiers may be employed.[3]

BENEFITS AND LIMITATIONS Transducers of less than 1 mm diameter (dynamic use) or less than 2 mm (combined static and dynamic use) are available.

COMMENTS [1]Sensitivities in the pressure sensing direction as low as 0.00004% of full range/g for the higher ranges with cross axis sensitivity of 1/5 of this value are quoted.
[2]Conditioned to selected bands of approximately 50°C. This arises because of the thermal nonlinearity of semiconductor devices.
[3]Some are complete with integral signal conditioning as part of the transducer.

SUPPLIERS ACT; DRU; END1; END2; ENT; KUL1; KUL2; PEN; SEN3; SEN4.

Pressure, fluids

PR8. Quartz resonator

OPERATING PRINCIPLE A quartz crystal, when suitably energised, will oscillate at its natural frequency. If this same crystal is then subject to stress in the correct axis this frequency will change.

This is exploited here by subjecting a crystal to pressure, the resulting deformation changing the natural frequency of the crystal.

By using an electronic feedback circuit to maintain the resonant oscillation the frequency output change is a direct measure of the pressure.

APPLICATIONS One of the most precise pressure techniques available. Uses include measuring small tide height changes in deep ocean conditions or as a calibration standard.

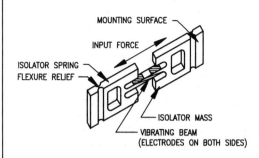

SINGLE BEAM LOAD–SENSITIVE QUARTZ RESONATOR

Courtesy Parascientific Inc

RANGE LIMITS 0 – 40 kPa to 0 – 69 MPa.

OVER-RANGE PROTECTION ×1.2 of full range.

ERROR Nonlinearity: ±0.005% of full range. Hysteresis: ±0.005% of full range. Repeatability: ±0.005% of full range.

SUPPLY VOLTAGE 6 V to 35 V dc supply to inbuilt signal conditioning.

SENSITIVITY Resolution of 1×10^{-8} of full range.

CROSS SENSITIVITY Acceleration sensitivity is low and a value of 0.0038% of full range/g is quoted.

TEMPERATURE RANGE Operating: 0°C to +125°C (calibrated range).

TEMPERATURE SENSITIVITY Temperature error is incorporated within the error values quoted above.

SIGNAL CONDITIONING Incorporated within the transducer; providing a waveform typically 4 V peak-to-peak square wave both for pressure and temperature.[1]

BENEFITS AND LIMITATIONS Allows measurements of very small pressure changes. A laboratory standard capable of use in the field.

COMMENTS Their precision will require the highest quality of calibration.
[1]Outputs may be further manipulated and enhanced by additional signal conditioning which will include the temperature correction and conversion of the signal into selected engineering units.

SUPPLIERS EXP; OCE.

Pressure, fluids

PR9. Remote capacitance transmitter

OPERATING PRINCIPLE With very many mechanically-operated pressure gauges, such as Bourdon gauges, already installed there are occasions when the information may be required remotely in an electrical form. It is possible to fit over the existing gauge a device which, when driven by the pointer movement, will result in some electrical change.
In the example given here a capacitive change is experienced and when measured gives an output proportional to the pointer pressure position.

APPLICATIONS Mounted over existing gauges, they are used anywhere that existing mechanical device measurement would be enhanced by remote data logging.

Courtesy Budenberg Gauge Co Ltd

RANGE LIMITS Above 35 kPa of pressure gauge full operating range.[1]

ERROR ±1% of full range.[2]

OVER-RANGE PROTECTION This is set by the pressure gauge over-range stop.

SUPPLY VOLTAGE 22 V to 36 V dc to inbuilt signal conditioning.

SENSITIVITY 4 – 20 mA or 0 – 20 mA for full range.

CROSS SENSITIVITY Insignificant other than potential transducer drag on the gauge pointer.

TEMPERATURE RANGE −25°C to +70°C.

TEMPERATURE SENSITIVITY Insignificant compared to the overall accuracy.

SIGNAL CONDITIONING Provided as a complete transducer system with a current loop output.

BENEFITS AND LIMITATIONS A simple and easy conversion although it does tend to inhibit reading the main dial unless viewed directly from the front.

COMMENTS A conveniently cheap way of signal conversion especially for existing gauge installations.
[1]This lower limit of operation is determined by the drive torque from the gauge pointer needed to overcome the transducer restraint whilst maintaining an accuracy to some acceptable level.
[2]To this must be added the pressure gauge accuracy. However, often the device may be tuned for both zero and range value and hence when calibrated *in situ* the total error can be less than the sum of the individual ones.

SUPPLIERS BUD.

PR10. Resistance potentiometer

OPERATING PRINCIPLE Pressure is applied to a diaphragm and the resulting movement is linked to the wiper of a resistance potentiometer.
The resultant voltage measured between the wiper contact and the potentiometer resistance is a direct measure of the applied pressure.

APPLICATIONS Well suited to a wide range of applications such as water level monitoring.
The relatively low power consumption is ideal for battery-operated data logging systems.

Courtesy Penny & Giles Ltd

RANGE LIMITS 0 – 20 kPa to 0 – 300 kPa.

ERROR Nonlinearity: ±0.5% of full range. Hysteresis: ±0.5% of full range. Repeatability: ±0.5% of full range.

OVER-RANGE PROTECTION ×1.25 without damage.

SUPPLY VOLTAGE 30 V dc is a likely maximum.[1]

SENSITIVITY The full range output signal is identical to the supply voltage.

CROSS SENSITIVITY Best under static or quasi-static conditions since they are likely to be sensitive to acceleration and vibration.

TEMPERATURE RANGE Operating: 0°C to +30°C. Storage: –30°C to +50°C.

TEMPERATURE SENSITIVITY Range: ±0.03% of full range/°C.

SIGNAL CONDITIONING Except for current loop systems no further amplification is required. The output may be via any suitable voltmeter.

BENEFITS AND LIMITATIONS The high output lends itself to any measurement of pressure where high accuracy or high frequency response is not essential.

COMMENTS [1]Lower voltages may be used to suit the required output or, if so desired, to match the numerical value of pressure. As an example 0 – 20 m water depth could be represented by 0 – 20 V. It is important that excessive current is not passed through the potentiometer and hence it is preferred that the output signal is input to a high impedance source.

SUPPLIERS PEN.

Pressure, fluids

PR11. Thick-film resistance strain gauge

OPERATING PRINCIPLE Essentially the technique is one of printing thick-film resistors and the associated "wiring" onto a ceramic disc .
These resistors are strain-sensitive and will respond to disc deformation typical of that caused by a pressure application.
This type of resistance or strain gauge has a gauge factor of the order of 10. They can be purchased as complete devices or as diaphragms alone for installation in a pressure measuring system.

APPLICATIONS These are well suited to many applications involving cus-tomised transducers for specific sys-tems.

Courtesy Sensit Ltd

RANGE LIMITS 0 – 10 kPa to 0 – 20,000 kPa.[1]

ERROR Nonlinearity, hysteresis and repeatability combined: between 0.4% and 0.2% of full range.[2]

OVER-RANGE PROTECTION Up to ×5 of full range subject to selection. Burst pressure ×2 of the highest operating range of a particular diaphragm.

SUPPLY VOLTAGE Bridge excitation voltage: up to 30 V dc.

SENSITIVITY From 0.06 mV/V/kPa to 0.0004 mV/V/kPa.

CROSS SENSITIVITY Will respond to accelerations due to mechanical vibration, the mag-nitude being dependent upon the choice of diaphragm.[3]

TEMPERATURE RANGE Compensated: –25°C to +85°C. Operating: –55°C to +125°C

TEMPERATURE SENSITIVITY Diaphragms alone: zero: $\pm 2\,\mu$V/V/°C (0 to +60°C); range: ±0.04% of full range/°C. Diaphragm with signal incorporated signal conditioning: zero: ±0.01% of full range/°C; range: ±0.01% of full range/°C.

SIGNAL CONDITIONING Digital voltmeters or dc amplifiers with a stable power supply may be used.[4]

BENEFITS AND LIMITATIONS Ceramic as a material is brittle so care should be taken when particle impact is likely at the pressure face.

COMMENTS [1]Each diaphragm, when offered separately, may cover a pressure range band, and could be used for example, between 0 – 100 kPa and 0 – 500 kPa.
[2]With the better performance at the higher pressure range values.
[3]Diaphragms must be mounted with sufficient clamping of the circumference to en-sure the type of bending experienced is consistent throughout the pressure range.
[4]Diaphragms may be purchased with the appropriate power and signal condition-ing, all incorporated on their "back face".

SUPPLIERS SEN2.

Pressure, fluids

PR12. Vibrating wire

OPERATING PRINCIPLE When plucked a pretensioned wire will vibrate at its natural frequency.[1]
If the wire is then subject to a change of tension the frequency will change. The wire is "plucked" by a pulse of dc voltage applied to an electromagnet, centrally located, and the resulting vibratory signal picked up by the same or a second electromagnet. The wire connects to a diaphragm.

APPLICATIONS Developed specifically for the civil engineering industry as used in soil mechanics studies. Three types are available, a pressure/load device (see also F09), pressure alone or a piezometer.

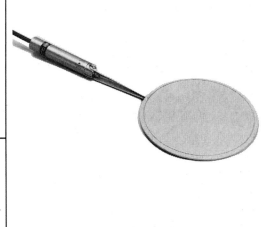

Courtesy Soil Instruments Ltd

RANGE LIMITS Pressure/load: 0 – 0.7 MPa to 0 – 2.5 MPa. Pressure: 0 – 5 MPa (oil filled); 0 – 20 MPa (mercury filled). Piezometer: 0 – 0.5 MPa to 0 – 10 MPa.

ERROR Nonlinearity and hysteresis. Pressure/load: ±0.5% of full range. Pressure: low pressure ±2 kPa of reading; high pressure ±0.1% of full range. Piezometer: ±0.1% of full range.

OVER-RANGE PROTECTION Dependent upon the design but some over-pressure can be tolerated without affecting the general performance.

SUPPLY VOLTAGE Self-generating but a voltage of between 10 and 50 V dc is used to "pluck" the wire.

SENSITIVITY Dependent upon the ability to measure changes of frequency.

CROSS SENSITIVITY Pressure alone must be applied to the diaphragm(s). For example, for soil studies, local small stones must not contact the plate surfaces.

TEMPERATURE RANGE Typical ambient temperatures.

TEMPERATURE SENSITIVITY Low, with the wire having the same coefficient of expansion as the body of the load cell.[2]

SIGNAL CONDITIONING Purpose-built; the display provides a linear correlation between load and display value. This same unit will provide the "plucking" voltage.

BENEFITS AND LIMITATIONS Purpose-built, for high stability long-term use. Because these are frequency-controlled devices, long cable leads may be used.

COMMENTS [1]The relationship is: Frequency = $(t.g)^{0.5}/2L_w.\sigma$ where t = tension, g = gravity, L_w = wire length and σ = density of wire.
[2]It is possible to include a thermometer to correct residual error.

SUPPLIERS SOI.

Pressure, Sound

Although the vast majority of microphones are used for speech or musical reproduction there is a significant demand for the measurement of other types of air- or water-borne pressure waves. This section describes both types of measurement, identified as "audio" or "sound intensity" respectively.

Microphones are pressure-measuring devices which operate over a wide frequency band and generally very low pressure levels. Some devices are sold on a dual basis, that is both for fluid pressure and sound measurement. Being developed as specialist devices used to convert airborne pressure into electrical signals they have evolved their own technology and phraseology within the wider audio market. It is not within the scope of this book to discuss this in great length. However some of the terms are useful in gaining an insight to microphone performance. The more common are as follows:

1. The decibel. The unit is the "Bel" named after Alexander Graham Bell but the decibel, 1/10th of the Bel, is preferred since it is more conveniently sized.

The decibel is a measure of the power ratio such that two powers, W_1 and W_2, may be compared as follows:

$$N = 10\log_{10}(W_1/W_2)$$

where N is the power ratio measured in decibels.

In sound measurement N is known as the **sound power level** (PWL); not to be confused with sound pressure level (SPL) discussed below.

Without delving too far into the use of the decibel it is of direct relevance to human hearing which is logarithmic in its response to sound intensity.

It is important to remember that the Bel is a ratio and is related to a chosen datum which, for sound measurement, is the threshold of hearing termed zero decibels. Sound levels below this level will have a negative value.

It is more convenient to measure voltage or current in audio circuits in terms of "gain" in decibels and then it is necessary to modify the above expression when using the decibel for such an electrical measure. If say the input and output impedance of an electrical amplifier are equal the voltage or current ratio becomes:

$$W_1/W_2 = (V_1/V_2)^2 = (I_1/I_2)^2$$

Then for the voltage ratio:

$$N = 10\log_{10}(V_1/V_2)^2 = 20\log_{10}(V_1/V_2)$$

and similarly for the current ratio.

It is unfortunate that the voltage (or current) ratio is sometimes misused to describe a power ratio and should be avoided. The relationship between the two are that the voltage ratio is the square root of the power ratio value.

Table 1. Typical values of power and voltage ratios in decibels

dB	Power ratio	Voltage ratio
1	1.259	1.122
10	10.0	3.162
50	100,000	316.2
100	$10,000 \times 10^6$	10^5

2. Sound Pressure Level In a similar way the sound engineer prefers sound pressure rather than power and hence sound pressure level (dB SPL) is used, conveniently related to the threshold of hearing. Because, as for the voltage case above, the sound power is proportional to the (sound pressure)^2 then:

$$\text{dB SPL} = 20\log_{10}\frac{(Measured\_sound\_pressure)}{(Threshold\_of\_hearing)}$$

Note. The value of "the threshold of hearing" (0 dB SPL) was established by an experiment on a group of young people and set at 0.00002 Pa (rms). The "threshold of pain" to the human ear has a sound pressure of 20 Pa or a dB SPL of 120.

3. Sensitivity In microphone transducer technology it is also convenient to consider the transducer sensitivity in convenient electrical terms. This may be expressed as:

$$\text{Sensitivity} = 20\log_{10}\frac{(Output\_Vrms\_at\_1\_Pa)}{(1\_Vrms\_at\_1\_Pa)} \text{ per Pa}$$

This sensitivity is quoted in dB with reference to 1 V/Pa or simply dBV. It is not uncommon for other units to be used such as mV/Pa or mV/μbar. Note that it is also convenient to identify a microphone's sensitivity to vibration to its "equivalent SPL" at 1 g.

4. Equivalent noise level. This is the electrical noise self-generated by a microphone compared to 0 dB SPL.

5. Directional characteristics. This is a measure of a microphone's directional or polar response to sound waves.

There are three main response patterns or polar diagrams and these are:

a) Omni-directional, where the directional sensitivity is generally uniform throughout 360°.

b) Bi-directional, where the polar diagram approximates a figure-of-eight identifying one axis of maximum sensitivity and the mutually perpendicular axis of minimum sensitivity.

c) Uni-directional, where the microphone has a maximum sensitivity from one direction alone. Because the polar diagram tends to be heart-shaped, these devices may be identified as **cardioid** (see Fig. 2) or hypercardioid if the directional sensitivity is very narrow.

6. Frequency response. This may be expressed numerically by identifying the lower and upper frequencies of use, which preferably should be accompanied by an indication of the amplitude linearity or by a graphical representation where the total characteristic is given. A typical graph is shown in Fig.3.

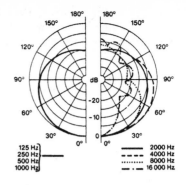

Figure 2. *Typical cardioid-shaped polar response curve*

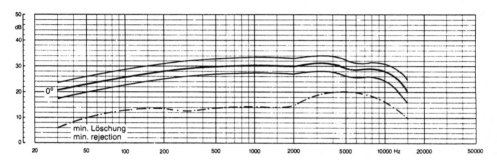

Figure 3. Typical microphone frequency response (Figs 2 & 3 courtesy AKG Acoustics Ltd)

To a measurement engineer it seems sacrilege for a transducer response to be modified so that it will deliberately distort some part of the measurement but this is commonly the case for audio microphones. The "enhancement" may involve suppression of the low frequency resonances, which might occur in large halls, or an increase in sensitivity at the middle of the frequency range to improve the timbre of instrument or voice.

Calibration is essentially a dedicated activity and can not be attempted without both the appropriate facilities and expertise.

NAMAS Accredited Laboratories: 0144 and 0237.

Contents: Pressure, sound

PS1. Audio, capacitance microphone

Sound pressure level: 132 dB SPL to 142 dB SPL. Frequency response: 20 Hz to 20 kHz.

PS2. Audio, dynamic microphone

Sound pressure level: 124 dB SPL to 128 db SPL. Frequency response: 20 Hz to 20 kHz.

PS3. Sound intensity, piezo-electric

Sound pressure level: In excess of 180 dB SPL. Frequency response: 1 Hz to 10 kHz.

PS4. Sound intensity, piezo-resistive

Sound pressure level: 100 dB SPL to 190 dB SPL. Frequency response: zero to 20 kHz.

Pressure, sound

PS1.   Audio, capacitance microphone

OPERATING PRINCIPLE A typical sensor element comprises a plate which forms one half of a capacitor; hence it is usually known as a condenser microphone.

This plate or diaphragm experiences a vibration when a variable air pressure is applied and the resulting change of capacitance is detected and converted into an equivalent voltage change.

A double diaphragm may also be used for omni directional detection. They come in various design options. This type of microphone will contain all of the essential electronics within the microphone body.

APPLICATIONS Used for faithful reproduction of audible sound, these microphones are also used for detecting and recording noise in machinery, etc.

Courtesy RS Components Ltd

RANGE LIMITS Sound pressure level: 132 dB SPL to 142 dB SPL.[1]
Frequency response: 20 Hz to 20 kHz.

ERROR Frequency response: ±1 dB from maker's supplied curve.[2]

OVER-RANGE PROTECTION Over-ranging will result in distortion of the output. 0.5% total harmonic distortion is typical.

SUPPLY VOLTAGE A power supply between 9 V to 52 V is typical.

SENSITIVITY –64 dBV at a frequency of 1 kHz. (6 mV/Pa approximately).

CROSS SENSITIVITY Equivalent noise level: 15 dB SPL. Crosstalk rejection: better than 70 dB (20 Hz to 10 kHz); better than 40 dB (20 Hz to 15 kHz).

TEMPERATURE RANGE Operating: –20 to +70°C

TEMPERATURE SENSITIVITY Insignificant.

SIGNAL CONDITIONING Separate supply and further amplification connected to the microphone by phantom powering, allowing the transmission of the supply and return signal by only two wires.

BENEFITS AND LIMITATIONS A very common microphone used for a wide variety of speech and musical sounds.

COMMENTS [1]The lowest response to sound pressure is influenced by factors described under Cross Sensitivity.
[2]The frequency response may be deliberately modified from a constant value as discussed in the introduction to this section.

SUPPLIERS AKG1; AKG2; RSC.

Pressure, sound

PS2. Audio, dynamic microphone

OPERATING PRINCIPLE There are a number of design styles for dynamic microphones. In one form a diaphragm, to which is attached a lightweight inductive coil, detects the sound pressure waves.
The coil in turn is surrounded by a strong radial magnetic field. Hence the vibratory motion of the coil generates an electrical signal. Alternatively, a central lightweight aluminium rod may be attached to the diaphragm and its motion through the coil provides the signal.

APPLICATIONS Perhaps the most common type of all microphone options with a variety of uses such as portable tape recorders or public address systems.

Courtesy RS Components Ltd

RANGE LIMITS	Sound pressure level: 124 dB SPL to 128 db SPL. Frequency response: 20 Hz to 20 kHz.
ERROR	Dependent upon the particular design but generally less linear than capacitance microphones.[1]
OVER-RANGE PROTECTION	Over-ranging will result in distortion of the output. 1.0% total harmonic distortion is typical.
SUPPLY VOLTAGE	Self-generating
SENSITIVITY	In the order of –76 dBV at 1 kHz (1.6 mV/Pa approximately).
CROSS SENSITIVITY	Hum sensitivity[2]: at 50 Hz, 3 μV/5μTesla; at 100 Hz, 6 μV/5 μTesla.[3]
TEMPERATURE RANGE	Operating: –10°C to +65°C
TEMPERATURE SENSITIVITY	Insignificant.
SIGNAL CONDITIONING	Although these devices are self-generating further amplification is essential. Various options such as suitable filters are available.
BENEFITS AND LIMITATIONS	A good competitively priced product with many options and variations.
COMMENTS	[1]This distortion however, may be used to advantage in many sound applications. [2]This type of microphone is prone to interference from mains frequencies. To counter this "hum coils" are used to provide a compensating voltage and result in the quoted values. [3]The Tesla is a measure of magnetic flux density or "hum field". 5 μTesla is approximately equal to 50 mGauss.
SUPPLIERS	AKG1; AKG2; RSC.

PS3. Sound intensity, piezo-electric

OPERATING PRINCIPLE A piezo-electric crystal exhibits the property that when deformed it yields an electrical charge, measured in Coulombs, proportional to that deformation. Thus by making the crystal part of a pressure sensitive diaphragm, sound waves will be detected and converted into an electrical signal.

APPLICATIONS These devices are most likely to be used in hazardous environments or where they are prone to excessive handling, such as in aircraft engine bays or in under-water applications.

Courtesy Endevco (UK)

RANGE LIMITS Sound pressure level: in excess of 180 dB SPL[1]
Frequency response: 1 Hz to 10 kHz.

OVER-RANGE PROTECTION Over-ranging will result in distortion of the output. Up to 5% total harmonic distortion may occur at 160 dB SPL.[2]

ERROR Nonlinearity, amplitude: ±0.5 dB from 120 to 164 dB rising to ±1.0 dB up to 180 dB is a typical performance. Frequency response: ±1 dB from 2 Hz to 4 kHz increasing to ±3 dB up to 10 kHz.

SUPPLY VOLTAGE Self-generating

SENSITIVITY 23 pC at 140 dB SPL equivalent to approximately 0.12 pC/Pa.

CROSS SENSITIVITY Vibration: 105 dB SPL max for 1 g peak up to 3 kHz.

TEMPERATURE RANGE Operating: –54°C to +260°C

TEMPERATURE SENSITIVITY Better than ±1 dB over the total temperature range.

SIGNAL CONDITIONING Suitable charge amplifiers are required.

BENEFITS AND LIMITATIONS Being piezo-electric devices they can withstand considerable change in static pressure without affecting the performance.

COMMENTS [1]With the lower limit determined mainly by the quality of the amplification available.
[2]These devices are particularly robust and can tolerate mechanical impact up to shock of 500 g and vibration of 150 g.

SUPPLIERS BRU1; BRU2; BRU3; END1; END2.

PS4. Sound intensity, piezo-resistive

OPERATING PRINCIPLE This type of device is related to the common piezo-resistive pressure transducers.
A silicon diaphragm has diffused into it a resistance strain gauge bridge. When energised these gauges respond to the diaphragm deformation, providing a pressure-related output.

APPLICATIONS Apart from specialised audio applications there are other specialist uses for detecting aircraft engine or marine propeller noise. Particularly suited for high intensity and low frequency applications. One such example is boundary layer effects on aircraft surfaces.

Courtesy Endevco (UK)

RANGE LIMITS Sound pressure level: 100 to 190 dB SPL or better. Frequency range: zero to 20 kHz

ERROR Nonlinearity, amplitude: ±0.5 dB from 100 to 190 dB SPL.
Frequency response: ±0.5 dB, 0 to 5 kHz, increasing to ±5 dB up to 20 kHz.

OVER-RANGE PROTECTION Diaphragm burst pressure: better than 197 dB SPL.[1]

SUPPLY VOLTAGE Bridge voltage, typically 15 V dc.[2]

SENSITIVITY −93 dBV equivalent to approximately 0.022 mV/Pa.

CROSS SENSITIVITY Vibration: 100 dB SPL max for 1 g rms over the operating range.

TEMPERATURE RANGE Compensated: −20°C to +100°C. Operating: −54°C to +107°C.

TEMPERATURE SENSITIVITY Extremely low over compensated range.

SIGNAL CONDITIONING High quality dc amplifiers of the required frequency response characteristics are essential.

BENEFITS AND LIMITATIONS Although operating from zero frequency (zero gauge pressure), this ability may inhibit its use where the quasi-static pressure variation is large.

COMMENTS [1]They are capable of withstanding very high mechanical impact shocks of 10,000 g and vibration to 500 g.
[2]Like all piezo-resistive devices the performance is influenced by the chosen voltage supply. Other voltages should only be used when so recommended by the manufacturers.

SUPPLIERS END1; END2; KUL.

Relative humidity

Before considering the techniques of measurement of humidity, or more properly relative humidity, it is necessary to understand some of the related terms.

Humidity is a measurement associated with the moisture content held in suspension by the atmosphere. The amount of water vapour in a unit mass of air at any moment is known as the humidity.

Warm air can hold more water vapour than cold and when air holds its maximum of water vapour at a given temperature the air is said to be saturated.

This leads to the term "relative humidity" which is the proportion of water vapour in the air at a given time when compared to the total amount of moisture necessary to achieve saturation at the same temperature. It is expressed as a percentage and abbreviated to RH.

The earliest method of measurement of relative humidity was by the use of two thermometers such as mercury-in-glass. One of the thermometer bulbs is covered in muslin and supplied with water via a cotton wick from a reservoir. This thermometer is called the wet bulb thermometer and the uncovered one the dry bulb thermometer, the whole being a **wet and dry bulb thermometer** or **psychrometer**. (An earlier name was a **hygrometer**.) It is normal to use the term psychrometer only for these wet and dry devices and this will be adhered to here.

In operation, the water around the wet bulb will evaporate at a rate dependent upon the dryness of the air resulting in heat loss with a corresponding drop in temperature. The difference in temperature of the two thermometers is a measure of the relative humidity. As an example of the wet bulb temperature differences at 20°C, 100% RH has a zero difference whereas for 56% RH the temperature difference will be 5°C.

For wet bulb temperatures below 0°C it is essential that the wet bulb is covered in ice and even so there is a discontinuity in the manner in which the temperature changes with relative humidity at the transition point. Note that the air temperature may be above freezing when icing occurs because of the cooling effect of evaporation.

When air reaches saturation at any given temperature this is known as the **dew point**, so named because such saturation causes condensation forming dew on any ground surface. If the dew point is below freezing the resultant condensate freezes and is known as **hoar frost**.

Some technical specifications will quote an operating relative humidity and occasionally qualify the reading with the phrase "noncondensing". This means that no water droplets can be allowed to form on the component parts.

Calibration is best achieved by comparison with some type of wet and dry bulb thermometer. They may be mercury-in-glass or electrical devices such as platinum resistance thermometers but care must be taken that the water evaporation is properly controlled.

Table 2. Relative humidity (per cent)

Dry bulb °C	Depression of wet bulb, °C		
	0	2.5	5
0	100	52	—
10	100	69	38
20	100	77	56
25	100	80	62
30	100	82	65

NAMAS Accredited Laboratories: 0060, 0072 and 0266.

Contents: Relative humidity

RH1. Capacitance, humidity meter

0 to 100% relative humidity.

RH2. Platinum resistance, psychrometer

0 to 100% relative humidity.

Relative humidity

RH1. Capacitance, humidity meter

OPERATING PRINCIPLE A capacitance is formed by two electrically conducting plates separated by a dielectric material which can be permeated by the local air.

Thus as the air, containing water vapour, enters this space there is a change in the dielectric value giving rise to a capacitive change.

An extremely fine filter surrounds the capacitor to protect against the ingress of dust particles.

APPLICATIONS Being self-contained and often portable, they are used for the measurement of humidity such as rooms or storage compartments but not usually the natural environment.

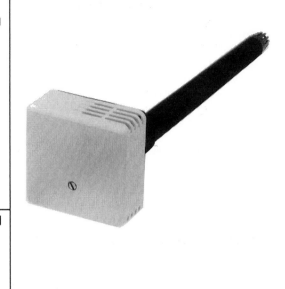

Courtesy RS Components Ltd

RANGE LIMITS 0 to 100% relative humidity.
ERROR ±2% RH is common. ±1% RH with calibration.[1]
OVER-RANGE PROTECTION None required since it operates over the full range.
SUPPLY VOLTAGE From low voltage battery operation to normal mains supply.
SENSITIVITY Current outputs: 4 to 20 mA.[2] Voltage output: 0 to 5 V.[2]
CROSS SENSITIVITY No water droplets must reach the sensing probe. Any filter must be clean to ensure air circulation.
TEMPERATURE RANGE Operating and storage: −40°C to +80°C.
TEMPERATURE SENSITIVITY ±0.04% RH/°C
SIGNAL CONDITIONING Usually part of a system. The probe is rigidly attached for portable use or remote as for duct or pipe installation.
BENEFITS AND LIMITATIONS Unlike true psychrometers which measure relative humidity directly, these devices require a calibration reference to establish their response.
COMMENTS [1]Quoted at a single temperature, normally 20°C. [2]Typical full range output.
SUPPLIERS EUR1; RSC; TES; VAI1; VAI2.

Relative humidity

RH2. Platinum resistance, psychrometer

OPERATING PRINCIPLE Two platinum resistance thermometers are used, one being "dry" and the other "wet", the latter maintained from a water reservoir. The resistance change of the two thermometers are both a measure of humidity and air temperature.

To ensure reasonable response a fan is used to pass air over the thermometer elements with a typical velocity of 4 m/s.

APPLICATIONS For precision measurement of relative humidity as is required for environmental monitoring and control.

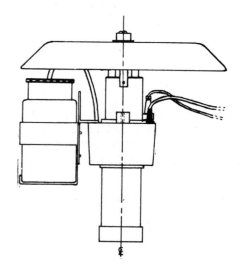

Courtesy Vector Instruments

RANGE LIMITS 0 to 100% Relative Humidity.[1]

ERROR ±0.2°C after calibration with a corresponding relative humidity error.

OVER-RANGE PROTECTION Being used for environmental measurements they can operate in wind speeds of up to 75 m/s.

SUPPLY VOLTAGE A typical input supply is 12 V dc.[2]

SENSITIVITY The temperature coefficient is 3.85 Ω/°C.

CROSS SENSITIVITY Care must be taken to ensure that the water container is maintained full and the thermometers are protected from solar radiation.

TEMPERATURE RANGE −20°C to +65°C

TEMPERATURE SENSITIVITY Not applicable: see Benefits and Limitations.

SIGNAL CONDITIONING This will require some form of bridge completion unit, a suitable bridge excitation supply and power for the fan.

BENEFITS AND LIMITATIONS By measuring RH in the classical manner, the conversion is simple and calibration is achieved by temperature readings alone.

COMMENTS A good example of a way in which a classical solution may be updated. The device provides a measurement of air temperature in addition to relative humidity.
[1]The thermometers have individual resistances of 100 Ω ±0.033 Ω at 0°C.
[2]The resistance elements will be operated at some small voltage supplied by the signal conditioning with additional power for the air circulation fan.

SUPPLIERS VEC.

Temperature and heat

Although "heat" is the fundamental quantity and "temperature" merely one of its manifestations, in practice temperature is the quantity which is measured for both heat and radiation when using electrical transducers. It is for this reason that in the general title to this section temperature receives priority.

The sensitivity to the effects of temperature of virtually all transducers would suggest that to measure temperature would be an easy task. Indeed there are a wide range of transducers to meet most needs in modern science and engineering.

However it is important to realise that temperature measured by **thermometers** will not necessarily give the desired requirement. Thermometers have a response time which may be too slow and hence the thermometer may never reach the desired level particularly in a continuously changing environment.

Nevertheless, under steady conditions the degree of accuracy can be very high. Ultimately accuracy is to a large extent limited by the ability of National Standards Laboratories to identify the "triple points" of various liquids to use as the reference points for the desired range of temperatures.

The first measure of temperature in any organised way was the Fahrenheit scale which is still widely used over a large part of the World. Its peculiar scale of 32°F and 212°F for the freezing and boiling points of water arises because the "zero" point was taken as that of the freezing point of a brine solution and 100° as the human body temperature; a reflection of things of import at the time of devising the scale and the quality of the calibration techniques!

Originally the measurement of temperature was based upon the expansion of a liquid such as mercury-in-glass thermometers. With the need for remote or continuous sensing, electrical devices were introduced. The most common of these are the **platinum resistance thermometer**, the **thermistor** and the **thermocouple** with others such as ultrasonic devices having specialist applications.

Selection of a thermometer to meet a chosen application must be made with care. It must be remembered why the temperature is being identified. Often what is required is to identify a representation of temperature. Take the measurement of temperature of an oven. Since the measurement is likely to be at a single position it says nothing about the temperature gradient in the oven space. However experience will confirm satisfactory performance once that temperature is reached, a secret known to all good cooks! If the oven tolerates a temperature variation of say 10°C there is little point in selecting a sensor with an accuracy of 0.1°C.

Again if temperature difference is the desire, a thermocouple may be a better choice even if its absolute temperature accuracy is possibly lower than some other device since only one measurement is required.

Calibration equipment of various degrees of sophistication are available from manufacturers. However the relating of temperature to its appropriate temperature standards is much more complex than such a simple measurement would suggest. For example confirming the performance of a precision thermometer at 0°C and 100°C will take patience and care. Before proceeding, do discuss the problem with the manufacturer.

Fortunately some temperature devices do not require in-house calibration since their performance is established by knowing the device's physical attributes. For others the process of incorporating them into temperature correction circuits will be one of using the manufacturer's performance specification and then making minor corrections during installation and testing.

NAMAS Accredited Laboratories: 0013, 0031, 0033, 0034, 0043, 0056, 0058, 0072, 0076, 0093, 0100, 0104, 0107, 0112, 0120, 0123, 0128, 0133, 0134, 0137, 0139, 0166, 0171, 0175, 0180, 0200, 0202, 0208, 0234.

Contents: Temperature and heat

Temperature and heat

TE1. Infrared, noncontacting

OPERATING PRINCIPLE All objects emit energy at all temperatures above absolute zero, including those in the infrared frequency band.

By detecting this particular emission and concentrating the radiation on to a thermocouple, by use of a lens or mirror, a measure of temperature of the viewed body may be made. Because of the nature of the measurement some devices are capable of measuring temperature and heat loss (or gain) of surfaces.[1]

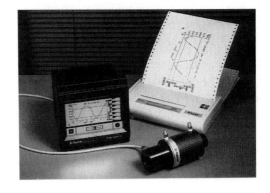

APPLICATIONS Specially designed for remote sensing such as furnace temperatures or where intrinsic safety may inhibit the use of electricity at the measurement station.

Courtesy Calex Instrumentation Ltd

RANGE LIMITS Temperature: –46°C to +3000°C in ranges.
Heat flow: 0 to 2 kW/m^2 from 0 to 260°C.

ERROR Accuracy: ±1% of reading, ±1 digit of display. Repeatability: ±0.5% of reading, ±1 digit.

OVER-RANGE PROTECTION Values are maxima but over-ranging causes no damage. For higher temperatures some sensing heads may be air or water cooled.

SUPPLY VOLTAGE Battery operated (portable) or mains power energised.

SENSITIVITY 1 mV/°C, typical.

CROSS SENSITIVITY The emissivity of different materials affects the performance unless corrected. These devices can provide correction within the signal conditioning.

SIGNAL CONDITIONING A choice of sophistication from complex systems to a device for converting a standard multimeter to give a temperature reading.[2]

BENEFITS AND LIMITATIONS A remote measurement allowing continuous monitoring without contact interference. The temperature is generally measured over a small area rather than a "point". The devices are simple to use being essentially "point and read".

COMMENTS [1]The subject of emissivity is complex and it is preferable to use different radiated wave lengths for different purposes.
[2]Ouputs can include current loop, digital (RS232) or voltage.

SUPPLIERS CAL; EUR1.

Temperature and heat

TE2. Resistance, bonded metal foil

OPERATING PRINCIPLE These devices have the superficial appearance of the bonded resistance strain gauge and comprise a grid of metal attached to a backing of insulating material, the whole being bonded to the surface of interest.

The metal element is selected to provide a high response to temperature and a low one to mechanical strain; that is their TCR is of a high value.[1] (See Strain for an explanation of TCR).

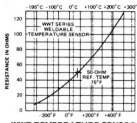

TYPICAL TEMPERATURE RESPONSE CURVE

WWT TEMPERATURE SENSOR

APPLICATIONS Convenient where strain gauges are also being used to measure on the same element, especially as bonding and wiring is achieved with the same available skills.

Courtesy Measurements Group (UK) Ltd

RANGE LIMITS −195°C to + 260°C.[2]

ERROR ±1% for full range or ±0.5% part range. Both after "shunting".[3]

OVER-RANGE PROTECTION Essentially the range is the limit of operation. However these devices are used most commonly over a portion of the range.

SUPPLY VOLTAGE 0.15 V across gauge or much higher if self-heating can be tolerated.

SENSITIVITY 0.3 Ω/°C typical and dependent upon foil type.

CROSS SENSITIVITY A small sensitivity to mechanical strain and smaller still to transverse strain.

SIGNAL CONDITIONING Dependent upon use but dc amplifiers or resistance measurement may be used.[4]

BENEFITS AND LIMITATIONS These devices are best for monitoring the parent material and not the environmental temperature.

COMMENTS [1]The type of metal of the foil will depend upon the application.
[2]"Life-limited" at the upper extremes of temperature.
[3]These devices are nonlinear with slightly different characteristics for different foils and parent materials. The linearity is improved by the use of a shunt resistor across the device, although reducing the available output.
[4]Take care that the resistance/temperature characteristics of the connecting cable does not produce any undesirable or unidentified errors.

SUPPLIERS MEA1; MEA2.

Temperature and heat

TE3. Resistance, bonded semiconductor

OPERATING PRINCIPLE These devices have the superficial appearance of the bonded semiconductor strain gauge discussed elsewhere in this book.

They comprise a single piece of "N" doped silicon material and are bonded directly to the surface of a parent body.

They are the complement of the semiconductor strain gauge in that the silicon element is chosen to provide a high response to temperature and a low one to mechanical strain.

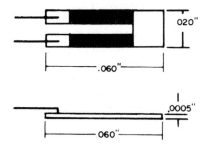

Courtesy Kulite Sensors Ltd

APPLICATIONS Well suited for a range of temperature measurements particularly for in-house constructed thermometers especially where high output is desired.

RANGE LIMITS −55°C to +250°C.

ERROR Nonlinearity: ±0.0045% of full range/°C.[1]

OVER-RANGE PROTECTION Although the range is the limit of designed operation, they are capable of operating at lower temperatures but are inherently nonlinear.

SUPPLY VOLTAGE Dependent upon requirement but typical Wheatstone bridge supplies are suitable.[2]

SENSITIVITY 2 mV/V of bridge supply/°C.

CROSS SENSITIVITY Thermal effects: 0.018% of full range/°C when bonded to material with a thermal coefficient of 18 μe/°C.[3]

SIGNAL CONDITIONING Since the output is so large, any suitable voltmeter such as a multimeter will suffice.

BENEFITS AND LIMITATIONS Extremely small devices, typically 2 mm long. With no backing carrier their temperature response should be good.

COMMENTS [1]Nonlinear as a simple resistance; particularly so for the lower temperatures (below 0°C). However, when part of a Wheatstone Bridge of suitable resistance value the linearity is as stated.
[2]There is some self-heating but being of high resistance (10,000 Ω at ambient temperature), the voltage across the bridge can be reasonably high.
[3]They respond to mechanical or thermal strain between the silicon and the parent material of which the quoted value is an example.

SUPPLIERS KUL1; KUL2; MIC3; SAN.

Temperature and heat

TE4. Resistance, platinum

OPERATING PRINCIPLE A high purity platinum wire is formed or wound into a strain-free resistance. One of the properties of platinum is that its electrical resistance change with temperature is sufficiently large to allow a high order of measurement accuracy. [1] The performance of a platinum resistance thermometer should obey a temperature/resistance change relationship such as defined by the International Practical Temperature Scale 1968 (PTS 68).

APPLICATIONS Because of the high precision, they are used for a very wide number of measurements from industrial process applications, water temperatures and meteorological applications.

Courtesy Rosemount Engineering Co Ltd

RANGE LIMITS −200°C to +800°C.

ERROR Accuracy: ±0.01°C after calibration. [2]

OVER-RANGE PROTECTION Range limits quoted are maxima. They are highly stable when subject to both long term extremes of temperature and thermal shock. [3]

SUPPLY VOLTAGE Sufficiently low to minimise self-heating.

SENSITIVITY 0.4 Ω/°C/100 Ω resistance.

CROSS SENSITIVITY Sensitive to induced mechanical strain on the platinum element. Usually made immune from high vibratory forces.

SIGNAL CONDITIONING Any form of resistance measuring equipment of sufficient accuracy is suitable.

BENEFITS AND LIMITATIONS Platinum resistance thermometers come in a wide range of sizes and shapes. Many of them are designed for specific industries. They are generally the most accurate of all of the electrical temperature measuring devices. Since the resistance element needs to be well protected their temperature response is not rapid.

COMMENTS Since these devices experience self-heating it is sometimes forgotten that in fluid flow they can suffer from a cooling effect.
[1] Values of the temperature coefficient of platinum are identified with great precision; the nominal temperature coefficient is 0.00385 Ω/Ω/°C.
[2] The change of the electrical resistance of platinum is slightly nonlinear. As an example at 0°C if the resistance value is 100.00 Ω, at 20°C the value is 107.79 Ω and at 40°C the value is 115.54 Ω.
[3] Within ±0.05% of the resistance value after subjected to 10 consecutive thermal shocks.

SUPPLIERS AUT; ABB2; LAB; RHO; ROS; RSC; TEC; VEC; VER.

Temperature and heat

TE5. Resistance, thermistor

OPERATING PRINCIPLE Thermistors today are part of modern silicon chip technology in which a small silicon junction is formed to give a significant change of resistance when subject to a change of temperature.

They are provided with positive or negative TCRs in a range of resistance values, typically: 2252, 3000, 5000, 10,000, 30,000, 50,000 and 100,000 Ω. (All values quoted at 25°C).

APPLICATIONS Many uses in a wide range of applications including air flow and temperatures measurements, temperature control systems as used in modern electronic equipment cooling and stabilisation systems.

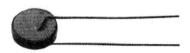

Courtesy RS Components Ltd

RANGE LIMITS −80°C to +150°C.

ERROR Accuracy: ±0.2°C, 0°C – 70°C (without calibration).

OVER-RANGE PROTECTION Essentially the range is the limit of operation. For most applications a much smaller range than that shown above is used.

SUPPLY VOLTAGE This will depend upon the way in which they are incorporated into a particular measurement circuit.

SENSITIVITY Example of resistance at 0, 10, 20 and 30°C steps are 7352.8, 4481.5, 2812.8 and 1814.4 Ω.

CROSS SENSITIVITY Electrical current flow through thermistors does produce self-heating and as an example 1 mW can raise its temperature 1°C.

SIGNAL CONDITIONING Commonly incorporated into part of the signal conditioning of some other type of transducer to provide measurement or correction for some unwanted temperature effect.

BENEFITS AND LIMITATIONS Their extreme nonlinearity is their biggest disadvantage for use as a thermometer but linearising circuits are produced for small temperature excursions.

COMMENTS Their temperature response is not rapid, typically 10 s to reach approximately 60% of the difference between its initial temperature and a new impressed one. (Whether this is a limitation or not depends upon the application). Their cost is low which is one of the reasons for their frequent inclusion as part of an assembled or integrated circuit.

SUPPLIERS ABB2; AIR; OCE; RHO; RSC; TES; VAL; VER.

Temperature and heat

TE6. Thermocouple

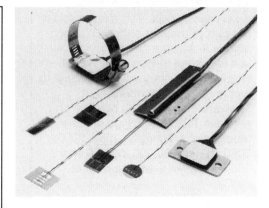

OPERATING PRINCIPLE If two wires, each of a different metal, are placed in intimate contact to form a "thermojunction", at one end and the free ends, the "cold junction", are held at a common temperature then when there is a temperature difference between the joined and free ends there will be an emf generated which is measurable at the free ends.

In modern devices the preferred choice of metal combinations and their performance is well documented.[1]

APPLICATIONS Very commonly used for applications such as furnaces, temporary installations and general monitoring and control systems.

Courtesy Rhopoint Ltd

RANGE LIMITS $-250°C$ to $+1300°C$.[2]

ERROR Class 1: $\pm0.5°C$ or $\pm0.004 \times$ the thermojunction temperature. Class 2: $\pm1°C$ or $\pm0.015 \times$ the thermojunction temperature. (Reference junction at $0°C$).[3]

OVER-RANGE PROTECTION The range is the limit of operation. The higher extremes of a particular range may be time-limited at that temperature.

SUPPLY VOLTAGE Self-generating.

SENSITIVITY Examples: Iron – Copper Nickel: 0.05 mV/°C ($0°C$). Nickel Chromium – Nickel Aluminium: 0.039 mV/°C ($0°$ C).

CROSS SENSITIVITY The thermoelectric effect is a function of the two wires and not just the thermojunctions; defined by the Law of Intermediate Temperatures. Thus, heating along the thermocouple wires may give rise to error.[4]

SIGNAL CONDITIONING A modern microvoltmeter is suitable. Electronic cold junctions are available to replace fixed temperature cold junctions.

BENEFITS AND LIMITATIONS Although of low output, they are well suited to measure temperature difference and have a high speed response to temperature change.

COMMENTS [1]The thermoelectric effect is well established and reported. However it is a complex subject. Appropriate textbook reading is of value. (See also the Bibliography).
[2]Dependent upon the chosen metallic combination and upon the manner in which the two wires are insulated.
[3]Thermocouples are identified into tolerance classes of which these are examples.
[4]If a third metal is inserted into the system, such as signal wires, there is no change in the total emf in a properly engineered system.

SUPPLIERS AUT; ENT; EUR1; LAB; RHO; ROS; RSC; TES.

Temperature and heat

TE7. Thermocouple, heat transfer.

OPERATING PRINCIPLE When heat flows from one location to another by conduction there is a resultant temperature differential.

Thus by placing thermometers across a preselected thin foil and attaching this package in intimate contact with the outer surface of a pipe, heat transfer can be measured across the foil, representative of the heat flow at the pipe wall. Temperatures are measured by a thermopile.[1,2]

APPLICATIONS Attached by bonding or clamping, allowing retrofitting, they are used for any application requiring study of heat loss or transfer.

Courtesy Rhopoint Ltd

RANGE LIMITS $0 - 113$ kW/m^2 to $0 - 567$ kW/m^2 (at 21°C).[3]

ERROR Dependent upon the type of use and the quality of surface contact.[4]

OVER-RANGE PROTECTION The limitations will be one of maximum temperature operation, as defined below, rather than heat flow.

SUPPLY VOLTAGE Self-energising.

SENSITIVITY Between 0.006 and 3.488 mV/W/m^2 at 21°C.[3]

CROSS SENSITIVITY Devices influence the heat transfer locally thus: thermal capacitance: 204 to 1021 J/m^2/°C; thermal impedance: 0.0020 to 0.0005°C/W/m^2

TEMPERATURE RANGE Operating: −200°C to +260°C.

SIGNAL CONDITIONING Any suitable microvoltmeter may be used but more sophisticated systems may be required for continuous monitoring.

BENEFITS AND LIMITATIONS A simple and reliable system for heat transfer problems requiring little skill or training to fit retrospectively.

COMMENTS [1]A thermopile is a number of thermocouples connected in series with each junction insulated from its neighbours by a thermal barrier.
[2]The overall dimensions may be described as "postage stamp" size with the thickness being between 0.076 mm and 0.330 mm dependent upon maximum heat flux and sensitivity.
[3]The performance varies in a nonlinear manner with temperature and suitable correction data exists.
[4]Devices may be individually calibrated against a master calorimeter.

SUPPLIERS RHO

Time interval

The measurement of time is one of great historical interest. The use of sundials as universal time keepers throughout antiquity was a method which needed a great understanding if local solar time was to be related to equal units of time or "mean time".

The initial attempts to create a self-contained time interval standard used the principle whereby any regular happening or event will, under ideal circumstances, take the same time to recur. The water clocks of the Greeks, the burning of a candle or the fall of sand through a restriction are classical examples.

Early clocks improved on this approach whereby the regular movement of an arm being driven backwards and forwards by a suitable mechanism provided a better time keeper but of course all of these devices are extremely sensitive to external influence such as vibration and temperature.

The invention of the pendulum clock at last provided a device which had at its heart a simple harmonic motion control mechanism and time keeping was freed from the grosser effects of external error sources. The introduction of springs as both the source of the harmonic motion and the power, reduced the physical size; "clocks" could become "watches".

The transition from the pendulum to an oscillating spring was a relatively easy step and time interval was measurable to very high orders of accuracy in a portable device. Indeed, when compared to many other types of transducer, mechanical clocks and watches are extremely precise. A clock which gains 5 minutes a day may be to us a poor timekeeper but has a total error of better than 0.35%.

It is interesting that the initial drive for a very accurate spring-based time keeper came from the needs of ship navigation where pendulum devices were useless.

All earlier clocks were either weight or spring driven. The first serious use of electricity provided its own frequency reference, namely the mains driven electric motor which locked in on the regularity of the frequency of the supply voltage. Suitable gearing then provided the second, minute and hour hand movements.

This is of no value for portable devices, although some were constructed which detected local mains frequency radiation, and better ways were sought to improve portable time pieces resulting in the tuning fork watch. By maintaining the vibration of a tiny tuning fork, its regularity of vibration was used to provide the frequency standard. These devices were popular but had a relatively short life span because the digital watch replaced them.

Modern clocks and watches have remarkable orders of accuracy. A good digital watch with say an error of 10 s/month has an accuracy better than 0.0005%!

If greater accuracy is required, this is available from National Time Standards through continuous radio transmissions such as from the Rugby Radio Transmitter in the UK.

Just the two examples of time measurement are given in the following pages; the tuning fork, and the "crystal" clock.

Contents: Time interval

TI1. Crystal clock
Frequency: 1 MHz to 24 MHz.
TI2. Tuning fork
Frequency: 200 Hz to 4 kHz.

Time interval

TI1. Crystal clock

OPERATING PRINCIPLE A piezo-electric crystal is energised to enable it to oscillate at its own natural frequency. This mechanical movement results in the crystal generating an ac voltage at that frequency.

This output provides the frequency base for the clock and by using divider circuits the output is divided into conventional time units from fractions of a second to year periods. Devices are available as timers which control to preset intervals of time.

APPLICATIONS Widespread. Now the most common method of time measurement and used for watches, clocks, computers and a wide variety of control systems.

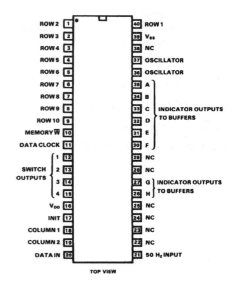

TOP VIEW

Courtesy RS Components Ltd

RANGE LIMITS Base frequency: 1 MHz to 24 MHz are commonly quoted.

ERROR 0.5 s/day is readily achievable at constant temperature.

OVER-RANGE PROTECTION Not applicable but care must be taken not to exceed the operating voltages recommended by the manufacturers.

SUPPLY VOLTAGE 1.2 to 1.9 V dc (liquid crystal display).
9 to 14 V dc (fluorescent display).

SENSITIVITY Possible to identify 0.01 s intervals, typical.

CROSS SENSITIVITY These devices are reasonably immune to normal shock and vibration conditions.

TEMPERATURE RANGE −40°C to +85°C typical.

TEMPERATURE SENSITIVITY ±0.0002% of the operating frequency/°C is common.

SIGNAL CONDITIONING This is incorporated within the "chip" and often contains additional features such as date and automatic leap year correction.

BENEFITS AND LIMITATIONS Digital displays for general time keeping is still resisted. Thus many watches and clocks still use "hands" for an analogue display, adding considerably to the cost.

COMMENTS The change from "clockwork" to the "silicon chip" for time pieces was a remarkable transformation. Within a decade the ability to measure time with precision was converted from one of laborious mechanical excellence to cheap mass production.

SUPPLIERS RSC.

TI2. Tuning fork

OPERATING PRINCIPLE An alloy steel tuning fork, located rigidly at the fork junction, has its tines vibrated at their natural frequency by inductive means.

The resulting output frequency is detected and used to maintain the drive frequency.

The output frequency may be used as an oscillator (sine wave) or as a timing element (square wave)

APPLICATIONS At one time these devices were particularly common. Now use is confined to replacement units and some special requirements where the crystal clock could not survive such as in high radiation areas.

Courtesy Adretta Ltd

RANGE LIMITS Preset frequency range of between 200 Hz and 4 kHz.

ERROR Normal setting accuracy: ±0.01% of frequency.

OVER-RANGE PROTECTION Not applicable.

SUPPLY VOLTAGE 5 V dc but higher voltages can be accommodated.

SENSITIVITY ±100 mV to ±300 mV peak to peak (sinusoidal).[1]

CROSS SENSITIVITY Attitude sensitivity: ±0.01% of set frequency, any attitude. Supply voltage: 0.01% of frequency/V.

TEMPERATURE RANGE −20°C to +70°C.

TEMPERATURE SENSITIVITY Operating: between −0.0002% and +0.0005% of set frequency/°C.

SIGNAL CONDITIONING These devices are supplied with all of the necessary circuitry to maintain oscillation.

BENEFITS AND LIMITATIONS These devices should be immune to high radioactive radiation.

COMMENTS The tines of the tuning fork are visible for applications which provide for isolated coupling to other systems by optical means.
[1]A square wave signal is also available of +500 mV for "0" and 5 V for "1"

SUPPLIERS ADR.

Torque

The need for precise torque application, particularly for manufacturing applications, is a relatively new activity in the field of measurement. Much of this need for control was initiated by the aircraft and the motor vehicle industries. In the past, control was influenced by practical training and experience where apprentices were instructed in spanner selection and the penalties of overtightening. As soon as safety requirements became dominant or where assembly was under mass production, controlled application of torque became essential.

Even so, as a measurement it has lagged behind its counterparts such as load or force and it is only in recent times that laboratories have existed which were capable of providing an accredited calibration service.

In considering torque transducers it is convenient to identify them in three groups namely:

1. Devices which apply torque, such as spanners and screw drivers. These applications are dominated by a wide range of mechanical devices.

2. Devices which both measure and transmit torque between the driver and driven, known as **torque meters**. These may be rotating or limited to a more static role; although it may require only that these latter devices be fitted with a signal transmission system such as slip rings to convert them into rotary ones. These devices may be used for calibration of other devices as described above. There are variations on this theme and examples include devices that may apply and measure both torque and tension for fastener performance studies, tests on child-resistant bottle tops or power tool run-down to simulate the tightening characteristics of fasteners. Within this group are included devices which are attached to existing transmission shafts and measure either strain or some other physical effect (identified in the text as "shipboard" since this is a common but not unique application). In such examples it is not possible to identify the overall accuracy since this will depend upon the shaft characteristics and so only the device performance is quoted. The ubiquitous bonded resistance strain gauge is commonly used and data transmission may be by slip rings or by some form of electrical transmission link. Each method has its admirers and here just one example is given. (The choice was dictated by showing an unusual technique not described elsewhere in this book.)

3. Devices which provide a calibration facility for the above. Torque is a "moment" and a good example is the torque produced by an electric motor. However, for most static applications the moment is usually achieved by the application of a known force at some chosen distance. This results in the transducer experiencing a further bending moment as well as the applied moment. Thus a transducer should be designed to be without response to this induced and unwanted moment.

For most applications, calibration devices exist but for the higher range of torque application, as experienced in heavy machinery, calibration is by some indirect method. This requires knowledge of the modulus of rigidity of the shaft material as well as details of its physical size; not always readily available on an existing installation.

193

There are variations of all of the three types of transducer listed above where the design is purely mechanical. Indeed for the most accurate calibrations the mechanical choice is by far the best. These are not listed here since the restrictions discussed at the start limits us to devices with an electrical output. However such mechanical devices do offer certain benefits and should always be considered when a choice has to be made.

Calibration of torque transducers depends upon the device. For torque spanners in the simplest manner the "nut" end of the spanner may be restrained in a vice and loads hung from the handle of the spanner.

For torque meters it is essential to build some form of static torque calibration rig. Essentially these rigs comprise an anchor point at one end and a torque-applying arm at the other. It is important that this arm allows weights to be hung from either end on scale pans and preferably the load should be transferred from one scale pan to the other rather than increase the weight from zero to the maximum at one end only. The reason for this is that in the former method the movement of weight produces a change in torque moment alone whereas the latter results in a change of bending moment on the transducer in the plane of the transducer rotational axis.

A mistake often perpetrated is to design the arm so that the horizontal axis of the two scale pan pivot bearings and the transducer attachment are in the same straight line although the outer bearings are a different size to the central one. It is important to maintain the actual point of pivoting in the same plane rather than the apparent rotational centre. The penalty for this error is that the system becomes over balanced and hence unstable, which may give rise to problems when establishing the zero stability.

NAMAS Accredited Laboratories: 0003 and 0256.

Contents: Torque

TO1. Torque meter, bonded resistance strain gauge (rotary)
±0.1 N.m to ±1000 N.m.

TO2. Torque meter, bonded resistance strain gauge (stationary)
±0.5 N.m to ±100 kN.m.

T03. Torque meter, piezo-electric
±200 N.m.

TO4. Torque meter, shipboard, bonded resistance strain gauge (rotary)
Unlimited.

TO5. Torque meter, shipboard, magnetostriction (rotary)
Unlimited.

Torque

TO1. Torque meter, bonded resistance strain gauge (rotary)

OPERATING PRINCIPLE Strain gauges are bonded to a shaft such that the gauges measure the principal strains resulting from the shear strain experienced by the shaft under torsional loading. (See also TO2.) Signals are then transmitted from the shaft by either slip rings or some other noncontacting method such as an inductive loop or radio frequency transmission.
Generally this decision is made by the manufacturer for the smaller sizes.

APPLICATIONS Most commonly mounted in line by flange mounting, they are used over a wide range of machine torque monitoring and control applications.

Courtesy Norbar Torque Tools Ltd

RANGE LIMITS ±0.1 N.m to ±1000 N.m in a wide number of ranges.

ERROR Nonlinearity and hysteresis combined: ±0.1% of full range.

OVER-RANGE PROTECTION ×1.5 without loss of performance.

SUPPLY VOLTAGE Subject to both the bridge resistance and the type of signal transfer technique but a maximum of 10 V is typical.

SENSITIVITY 2 to 3 mV for full range/V of input supply.

CROSS SENSITIVITY These devices are particularly insensitive to axial loads and any small bending errors will be averaged over one revolution.

TEMPERATURE RANGE Compensated: −10°C to +50°C Storage: −20°C to +70°C

TEMPERATURE SENSITIVITY Zero: 0.01% of full range/°C. Range: 0.03% of full range/°C.

SIGNAL CONDITIONING Decided by the method of signal transmission. For slip rings, a typical strain gauge amplifier system, dc or ac, is suitable.

BENEFITS AND LIMITATIONS A good linear system although it does require some form of signal transmission system which may introduce additional errors.

COMMENTS Possibly the most common method of measurement of torque on a rotating shaft. There are various preferences and indeed prejudices for the best ways of signal transmission. However a good rule is for small diameter shafts, silver slip rings with silver graphite brushes being the best and for larger shafts some form of noncontacting system is preferred including inductive loop and radio frequency transmission.

SUPPLIERS MHH; NOR; RDP.

Torque

TO2. Torque meter, bonded resistance strain gauge (stationary)

OPERATING PRINCIPLE Torsional loading of a shaft about its rotational axis results in a shear strain.
Since "shear" as such cannot be measured directly the resulting principal strains are measured, these being at an angle of 45 degrees to both the shear strain and shaft axis. Although the devices identified here are described as "static" many of them may be used for rotational applications subject to the addition of signal transmission equipment being fitted.

APPLICATIONS Installed in-line usually either by drive squares or flange mounting, they are used for a wide range of tasks from calibration to general torque measurement.

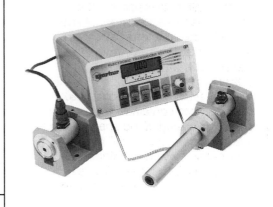

Courtesy Norbar Torque Tools Ltd

RANGE LIMITS ±0.5 N.m to ±100 kN.m.
ERROR Nonlinearity and hysteresis combined: ±0.1% of full range.
OVER-RANGE PROTECTION ×1.5 of full range.
SUPPLY VOLTAGE 15 V ac or dc maximum to strain gauge bridge.
SENSITIVITY 2 to 3 mV/V of input supply. 1 V for full range with incorporated signal conditioning.
CROSS SENSITIVITY Transverse sensitivity should be low with bending of the shaft likely to be the most dominant effect.
TEMPERATURE RANGE Compensated: −10°C to +50°C. Storage: −20°C to +70°C.
TEMPERATURE SENSITIVITY Zero: ±0.01% of full range/°C. Range: ±0.03% of full range/°C.
SIGNAL CONDITIONING Strain gauge amplifiers or digital voltmeters may be used.[1]
BENEFITS AND LIMITATIONS These devices are very reliable and have all of the general advantages of strain gauge transducers.
COMMENTS [1]Purpose-built equipment is available with a visual display in the appropriate engineering units often with the facility to display more than one type of unit combination such as "N.m" or "lbf. inch". Torque measuring devices have the potential to offer a better performance than force devices, particularly for non-linearity. However the present limitations are controlled by the capability of the calibration facilities which are generally of lower quality than force facilities.
SUPPLIERS DES; MEC; MHH; NOR; RDP; SEN1.

Torque

T03. Torque meter, piezo-electric

OPERATING PRINCIPLE A piezo-electric crystal exhibits the property that when a shear is applied across opposite faces of a piezo-electric disc or washer it yields an electrical charge, measured in Coulombs, proportional to that deformation.

A torque transducer will comprise a stack of discs contained within a protective casing. If the shear force is maintained the charge will gradually decline due to parasitic high resistance paths. Thus these devices are suitable for vibratory measurement alone.

APPLICATIONS Mounted so that the shear loading is transmitted by face friction, they are used wherever dynamic torque measurement is required, especially where a high static torque exists.

Courtesy Kistler Insturments Ltd

RANGE LIMITS ±200 N.m. (Partial range to ±20 N.m.[1])	
OVERLOAD PROTECTION ×1.2 of full range.	
ERROR Nonlinearity: ±1% of full range. Hysteresis: ±1% of full range.	
SUPPLY VOLTAGE Self-generating.	
SENSITIVITY 175 pC/N.m.	
CROSS SENSITIVITY ±0.0002 N.m/N of applied axial force.	
TEMPERATURE RANGE Operating: −196°C to +150°C.	
TEMPERATURE SENSITIVITY 0.02% of full range/°C.	
SIGNAL CONDITIONING These devices must be inputted to high impedance devices. Special amplifiers are available which convert the charge into voltage prior to further signal conditioning.	
BENEFITS AND LIMITATIONS This system has no comparable competitor. Although limited to dynamic measurement this can be an advantage since the output readings are not swamped with large static loads.	
COMMENTS [1]The "partial" range performance is claimed to be comparable to that of the full range values.	
SUPPLIERS KIS1; KIS2; KIS3; KIS4.	

Torque

TO4. Torque meter, shipboard, bonded resistance strain gauge (rotary)

OPERATING PRINCIPLE Strain gauges are bonded typically to the ship's propulsion shaft such that the gauges measure the principal strains resulting from the shear strain experienced by the shaft under torsional loading. There is a choice of signal transmission from simple slip rings to radio transmission systems.
The system identified here is one where the resulting analogue signal is translated into an ac pulse and is time modulated and then transmitted by some form of telemetry link.[1]

APPLICATIONS Retrofitted to existing transmission shafts it is commonly used on ships and other heavy machinery applications.

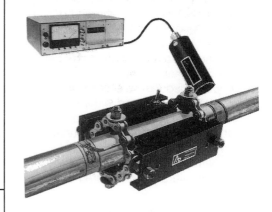

Courtesy Adretta Ltd

RANGE LIMITS Strain gauge bridge output: ±7.5 mV for full range.[2]
Bridge balance offset: ±2.5 mV.

ERROR Nonlinearity: ±3 μV of reading.

OVER-RANGE PROTECTION It is important to ensure that the clamp or other hold-down system can accommodate the rotational acceleration forces.

SUPPLY VOLTAGE Typical strain gauge bridge supply voltage is between 7.5 V and 30 V subject to the gauge resistance.

SENSITIVITY The period of the output wave form is typically 10 to 12 μs/mV input.

CROSS SENSITIVITY Strain gauge devices are particularly insensitive to axial loads and any small bending errors will be averaged over one revolution.

TEMPERATURE RANGE Shaft mounted components: 0 to +120°C.
Readout unit: 0 to +70°C.

TEMPERATURE SENSITIVITY Shaft mounted components: ±0.005% of full range/°C.
Readout unit: ±0.01% of full range/°C.

SIGNAL CONDITIONING The system comes in three parts: the shaft-mounted analogue converter and power supply unit, the transmission system and a read out unit.[3]

BENEFITS AND LIMITATIONS Pulse duration, whether for rotating or stationary applications provides a two-wire data transmission system.

COMMENTS [1]Note that the performance figures above are limited to the electronic systems alone and to this must be added the strain gauge errors and the uncertainty of the values chosen for the Modulus of Rigidity and shaft diameter. [2]Subject to the strain level available; in turn dictated by the power transmission through the shaft. [3]This last may incorporate an analogue or digital display with an electrical output for further data manipulation.

SUPPLIERS ADR.

Torque

TO5. Torque meter, shipboard, magnetostriction (rotary)

OPERATING PRINCIPLE The pattern of magnetic fields in a magnetic material is distorted by applying a mechanical stress field. Known as magnetostriction, it is used to measure the principal stresses created by torsion applied to a shaft. See also FO5. Three equispaced pole rings, holding a number of equispaced coils, are located around the shaft.
The magnetic coupling between rings is modified by the principal stresses and the magnetic anistrophy of the shaft, providing a measure of applied torque.[1]

APPLICATIONS Capable of retrofitting they are robust and ideally suited for shipboard and heavy machinery applications.

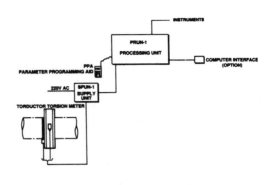

TORDUCTOR TORQUE AND POWER METER

Courtesy ABB Automation

RANGE LIMITS Stress level: 20 and 60 N/mm^2.
Strain level: 100 to 290 micro-strain for shipboard use.

ERROR Nonlinearity: ±1.0% of full range. Hysteresis: ±1% of full range. Repeatability: ±0.2% of full range.[2]

OVER-RANGE PROTECTION None required with little or no risk of damage.

SUPPLY VOLTAGE 110 to 440 V ac, 50 or 60 Hz.

SENSITIVITY Depends upon shaft stress but system full range output can be in the order of 10 V dc.

CROSS SENSITIVITY The system is particularly immune to both direct and bending stress fields.

TEMPERATURE RANGE Operating: 0 to +70°C.

TEMPERATURE SENSITIVITY Zero: ±0.01% of full range/°C Range: ±0.01% of full range/°C

SIGNAL CONDITIONING Supplied as part of a system, including a display. Added features are power readings (needing a tachometer) and energy consumption.

BENEFITS AND LIMITATIONS A particularly robust system. Requires but a short length of shaft for fitting. These devices must be tailored to pre-specified diameters.

COMMENTS [1]The pole rings are split, do not touch the shaft (separation 2.5 mm) and do not rotate. Thus the fitting of these devices is relatively simple and requires no shaft preparation, as with other types of ship shaft systems.
[2]For measurement system alone. To this must be added the uncertainty of the Modulus of Rigidity and the shaft dimensions.

SUPPLIERS ABB2; ABB3; INT1.

Velocity

Velocity is a measurement which the human being is incapable of experiencing. Unlike acceleration, force, pressure, temperature, displacement and many other measurements identified in the general text, velocity is never "experienced". When one considers our local velocity simply by living on the Earth's surface coupled with a combination of other velocities as we travel through space perhaps we should be grateful for this lack of feeling. However this inhibited early engineering ideas associated with velocity; as distinct from distance per unit time, the true concept of velocity did not materialise until mathematical precision defined the various relationships between displacement, velocity and acceleration.

It has been felt convenient in this book to consider the measurement of velocity of solids separately to that for fluids which are discussed elsewhere. This arises because of the nature of such measurement, where the flow of fluids is often associated with volumetric or mass flow measurement rather than displacement per unit time. Indeed it is arguable that the velocity of a solid body is more closely related to acceleration or displacement than it is to fluid flow; certainly in the manner in which these measurements are obtained and used.

Today, the measurement of velocity is a widespread activity (at least for the nonvector part of the quantity, that of **speed**). Speed is part of our commonplace experience. However there are only two electrical transducer techniques which measure speed directly without recourse to some more indirect effect. These are the voltage which is generated by the velocity of a body passing through a magnetic field, and the Doppler effect.

One example is given in this section with the Doppler technique presented under fluid flow. Although it would appear that these two techniques would be the most likely candidates for velocity measurement transducers this is not the case. Mechanical rotors are the preferred technique for fluids with a range of other devices in support and which are used for specialist applications. For solid body motion a combined measurement of displacement and time is often preferred. Indeed with modern techniques of precise time sampling techniques it is possible to adapt a wide range of displacement devices to a velocity measurement role, although none are presented here.

Velocity, angular

Like its linear counterpart, angular velocity can be measured by recording the displacement/time relationship of the output of any angular displacement transducer. Indeed under the proper circumstances a linear device may be used with effect. However to most engineers, the measurement of angular velocity is more commonly associated with continuous rotation as with shaft speeds.

Prior to the development of modern devices, belt-, gear- or chain-driven dc generators were probably the first electrical output transducers. Although still in use there is a move to cheaper systems such as pulsed output devices which not only allow precise counting techniques to be adopted but also identify shaft position, a technique which allows the initiation of some other activity at a precise angular position, IC engine ignition triggering being such an example.

Other than for the special case of the rate gyro, only angular devices which are capable of continuous rotation are included in this section and even then only those specifically identified by the manufacturers as angular rate rather than angular displacement devices.

Calibration of angular velocity transducers will depend much upon the type. Pulsed output devices require no calibration whereas electrical generators do, hence the former can be used to calibrate the latter. Calibration requires a speed controlled rotor to which the transducer can be attached. Precision speed controlled motors are available commercially.

NAMAS Accredited Laboratories: None known.

Contents: Velocity, angular

VA1. Rate gyroscope
±6°/s to ±5000°/s.

VA2. Rotation, magnetic
0.035 m/s to 8 m/s (local equivalent linear velocity).

VA3. Rotation, tachogenerator, ac generation
Up to 10,000 rpm.

VA4. Rotation, tachogenerator, dc generation
3500 rpm to 10,000 rpm.

Velocity, angular

VA1. Rate gyroscope

OPERATING PRINCIPLE A gyroscope is rotated in a gimbal suspension. When experiencing angular motion the rotor attempts to remain at a fixed attitude in space.

Spring restraints attached between the gimbal and casing overcome this resistance resulting in the spring deforming in proportion to the angular rate of change of the rate gyroscope. The deformation may be detected by a variety of pick-offs including resistive, inductive and capacitive types.

APPLICATIONS Often used in sophisticated systems such as aircraft manoeuvring control but has many other uses such as for research into moving systems especially vehicles, aircraft and ships.

Courtesy Smiths Industries Ltd

RANGE LIMITS ±6°/s to ±5000°/s.

ERROR Linearity: ±0.01°/s. Repeatability: ±0.01% of full range at 100 °/s. Hysteresis: less than 0.01°/s

OVER-RANGE PROTECTION High for both shock and vibration. Often specified to a Military Specification requirement.

SUPPLY VOLTAGE Subject to design but usually supplied from a dedicated power supply requiring a low voltage dc input.

SENSITIVITY Resistance pick-offs: up to 5000 Ω. Others: Up to ±4 V ac or dc.

CROSS SENSITIVITY Linear acceleration: 0.05°/s/g. Angular acceleration: 0.0006°/s/°/s^2.

TEMPERATURE RANGE Compensated: –40°C to +70°C. Operating: –55°C to +71°C.

TEMPERATURE SENSITIVITY Compensated: 0.045%/°C. Uncompensated: 0.35%/°C.

SIGNAL CONDITIONING Not an inherent requirement since the available signal is usually sufficient for most applications. For ac output devices signal rectification may be necessary.

BENEFITS AND LIMITATIONS Requires no spatial reference to measure angular velocity. Devices are limited to some lower frequency of operation, in the order of 10 Hz.

COMMENTS A device common in military applications and research oriented projects but perhaps not as well known as it should be in some industries. It is advisable to understand the manner in which rate gyroscopes operate. Hence it is recommended that the appropriate literature is sought. See also the Bibliography.

SUPPLIERS HUM; IFO; SMI.

Velocity, angular

VA2. Rotation, magnetic

OPERATING PRINCIPLE A coil is wound around a magnetic core to form an inductance. If a magnetic material passes close to one end of the magnetic core a voltage is induced into the coil.

The voltage output of these devices is influenced by both the velocity and closeness of the passing metal to the core.

(The induced voltage is proportional to the velocity, however it is the pulse generation which is the basis of measurement).

APPLICATIONS Many and varied. An example is a gear wheel providing pulses each revolution equal to the number of teeth.

Courtesy Ranco Controls Ltd

RANGE LIMITS Voltage range: 100 mV peak-to-peak to 33 V peak-to-peak.[1]
Velocity range: 0.035 m/s to 8 m/s.[1]

OVER-RANGE PROTECTION Normally none required. Care must be taken to ensure that the gap between the sensor and rotating device is maintained precisely.

ERROR Only dependent upon the quality of the frequency counting.

SUPPLY VOLTAGE Self-generating.

SENSITIVITY It is possible to operate these devices down to 50 mV or less.

CROSS SENSITIVITY Care must be exercised to ensure that no extraneous voltage pickup is present.

TEMPERATURE RANGE −20°C to +120°C.

TEMPERATURE SENSITIVITY Being a pulse signal there will be no significant heating effects.

SIGNAL CONDITIONING Dependent upon the final requirement but some form of frequency counter is required.

BENEFITS AND LIMITATIONS An extremely simple type of device but is unsuitable for measurement from zero rotational speeds.

COMMENTS The pulse shape comprises a rising signal followed by a sudden shift from a maximum to a minimum value followed by a gradual return to zero. When counting pulses it is important to ensure that only positive or negative pulses are counted and not both.
[1]For mild steel with a minimum gap of 0.25 mm. The maximum output is for a local velocity of 8 m/s and the minimum local velocity is for a voltage output of 100 mV.

SUPPLIERS APP; DEU; RAN1; RAN2.

Velocity, angular

VA3. Rotation, tachogenerator, ac generation

OPERATING PRINCIPLE This type of tachogenerator comprises a stator and rotor, the rotation of which produces a three phase ac output. The resulting voltage, both frequency and amplitude, is directly proportional to the rotational velocity. Being brushless, these devices offer a long life.

APPLICATIONS Used for the monitoring of shaft rotational speed over a wide range of machinery applications. An alternative to VA4.

Courtesy Powertronic International

RANGE LIMITS Up to 10,000 rpm.[1]

OVER-RANGE PROTECTION None required, other than the limit imposed by the maximum rotational speed.

ERROR Nonlinearity: ±1% of full range.

SUPPLY VOLTAGE Self-generating.

SENSITIVITY No load output: 24.5 V three phase ac (66.6 Hz). Equivalent dc: 31.5 V dc after rectification.

CROSS SENSITIVITY There will be a limiting current to prevent undue heating and for the example above it is limited to 35 mA.

TEMPERATURE RANGE Operating: –30°C to + 100°C.

TEMPERATURE SENSITIVITY Temperature coefficient: 0.02% of full range/°C (no load).

SIGNAL CONDITIONING Dependent upon needs. The system can be by frequency count or full three wave rectification to provide a dc output.

BENEFITS AND LIMITATIONS In dc mode they are less accurate than the ac equivalent.

COMMENTS [1]Note that when the ac voltage is rectified to produce an equivalent dc then there will be a limit to the lowest rotational speed measurement capability, typically 30 rpm. Where frequency measurement is preferred, the lower limit will be influenced by the voltage output when compared to the general electrical noise.
[2]The frequency generation is dependent upon the number of poles of the rotor. This example is for 8 poles and 1000 rpm.

SUPPLIERS HUB; POW.

VA4. Rotation, tachogenerator, dc generation

OPERATING PRINCIPLE These devices are permanent magnet multipole dc generators with brushes used to "pick off" the voltage generated; this voltage being proportional to the rotational velocity.

For best performance the brushes are made from a silver/graphite material.

APPLICATIONS Used for the monitoring of shaft rotational speeds over a wide range of machinery applications but with greater precision than the alternative at VA3.

Courtesy Powertronic International

RANGE LIMITS 0 – 3500 rpm to 0 – 10,000 rpm.

ERROR Nonlinearity: ±0.15% of full range.

OVER-RANGE PROTECTION None required, other than the limit imposed by the maximum rotational speed above.[1]

SUPPLY VOLTAGE Self-generating.

SENSITIVITY No load voltage: up to 150 V dc per 1000 rpm.[2]

CROSS SENSITIVITY It is important not to draw excessive current from the tachometer.

TEMPERATURE RANGE Operating: –30°C to + 100°C.

TEMPERATURE SENSITIVITY Temperature coefficient, no load: 0.005% of full range/°C (with compensation).

SIGNAL CONDITIONING This will depend only upon the final requirement since the voltage available is so high.

BENEFITS AND LIMITATIONS Good accuracy with a wide speed range and high no-load output voltage.

COMMENTS [1]These devices, require regular brush changes. A typical figure is in excess of 10^9 revolutions although at 1000 rpm continuous use this is some 19 years between changes.
[2]A change of rotation direction results in a change of sign, positive or negative, of the output voltage.

SUPPLIERS HUB; POW.

Velocity, linear

Velocity, as discussed in the previous section, is measured directly by only two methods, magnetic induction and Doppler effects. However for extremely low velocities it may be convenient to determine the displacement and time intervals separately and compute the velocity. This is particularly convenient in light of modern day computing techniques and where "time" is so readily established to a very high order of accuracy. This allows all of the displacement techniques described elsewhere to be used. However this section will be limited to true velocity devices; at least as defined by the manufacturers.

Calibration of devices which measure linear velocity is not as simple as it would first seem. This arises because of the mechanical difficulties of producing a constant speed over a large part of a relatively small displacement. Since for most applications the velocity towards the ends of the stroke must be zero there is a logic in subjecting the device to simple harmonic motion, as can be produced by a crank, and examining the sinusoidal waveform to establish the calibration quality. Modern computer analysis programmes provide quite sophisticated wave form analysis techniques from real-time data.

NAMAS Accredited Laboratories: None known.

Contents: Velocity, linear

VL1. Inductive, free mounted

Displacement: 12.5 mm to 500 mm. Frequency: in excess of 2 kHz.

VL2. Inductive, spring/mass system

Displacement: ±2.5 mm. Frequency: 1 Hz to 2 kHz.

Velocity, linear

VL1. Inductive, free mounted

OPERATING PRINCIPLE This uses the principle of Faraday's law of induction.
Two identical coils are wound side by side upon a nonmagnetic former and connected electrically.
Running through the centre of the former is a small cylindrical permanent magnetic core.
If this magnetic core is now mechanically connected to a vibratory source the differential electrical output from the windings will be proportional to the linear velocity of the core.

APPLICATIONS Well-suited to study structural vibration problems especially where unimportant high frequency vibration would swamp the output signal from an accelerometer.

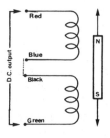

Courtesy RDP Group

RANGE LIMITS 0 – 12.5 mm to 0 – 500 mm in a range of displacement lengths or stroke.[1]

ERROR Nonlinearity: ±1% to ±5% of full range output.

OVER-RANGE PROTECTION The coils and core are not in mechanical contact. Hence in a well-engineered system, damage is unlikely.[2]

SUPPLY VOLTAGE None required

SENSITIVITY 6 mV/mm/s to 26 mV/mm/s (typical).

CROSS SENSITIVITY Insignificant.

TEMPERATURE RANGE Operating: –55°C to +95°C

TEMPERATURE SENSITIVITY Insignificant in terms of general performance.

SIGNAL CONDITIONING No amplification required. It is important that the measurement system is of high impedance to ensure no deterioration of the signal.[3]

BENEFITS AND LIMITATIONS These devices may be described as the "perfect" transducer, requiring no supply, giving a large output and responding directly to the desired physical quantity.

COMMENTS This type of transducer is often overlooked as a possible solution to a measurement problem.
[1]The frequency limit is generally restricted only by the voltage output limitations (see Note 3 below). It may be reasonably expected to operate in excess of 2 kHz.
[2]Where such contact may occur, low friction protective sleeves are supplied.
[3]The output voltage can be extremely high. One maker quotes the maximum velocity for any range is that which produces 500 V peak.

SUPPLIERS HBM1; HBM2; LUC1; LUC2; RDP.

Velocity, linear

VL2. Inductive spring/mass system

OPERATING PRINCIPLE This uses the principle of Faraday's law of induction. A coil is suspended in a magnetic field by two springs. When the whole is subjected to vibration the coil remains fixed in space and the resulting relative motion of the coil and magnet produces an electrical output across the coil equivalent to the applied velocity. The frequency of vibration should preferably be above the natural frequency of the spring/mass system.[1]

APPLICATIONS These devices are particularly suited for measurement on rotating machinery, particularly electric motors. Often used in hazardous environments.[2]

Courtesy HBM (UK) Ltd

RANGE LIMITS Displacement: ±2.5 mm. Frequency range: 10 Hz to 2 kHz extended down to 1 Hz with linearisation electronics.

ERROR Dependent upon design and application.[3]

OVER-RANGE PROTECTION High. In-bulit protection ensures easy handling of this transducer.

SUPPLY VOLTAGE None. Self-generating.

SENSITIVITY Between 0 – 4 mV/mm/s and 0 – 100 mV/mm/s dependent upon excursion range.

CROSS SENSITIVITY Transverse sensitivity is better than ±5% of full range dependent upon range.

TEMPERATURE RANGE −40°C to +90°C

TEMPERATURE SENSITIVITY Insignificant.

SIGNAL CONDITIONING Use the manufacturers equipment where the device is being used below its natural frequency; otherwise any suitable ac system may be used.

BENEFITS AND LIMITATIONS Available for horizontal use alone, vertical use alone or combined use; the last giving some degradation of frequency performance.

COMMENTS [1]The use of special electronics will allow quality measurement below the natural frequency.
[2]Such applications require properly approved devices to an accepted standard.
[3]Because of the nature of the output to frequency performance, particularly when near to the transducer natural frequency a value for nonlinearity would do little in describing this parameter. Further, the output is influenced by the amplifier input impedance and careful resistance matching can improve the linearity over a selected frequency range.

SUPPLIERS HBM1; HBM2.

In-house design

After World War II there was a remarkable upsurge in producing in-house transducer elements which are now commonplace purchases. Indeed this was the start of the electrical transducer industry where much of the research and development was initiated.

Strain gauges were constructed from wire, coils were wound for inductive devices and plates were put in close proximity to each other to use the capacitive effect.

At that time, in the 1950s, the mechanical and opto-mechanical devices reigned supreme. However the benefits of size, improved natural frequency, the capability to use a device more akin to the needs of the required measurement and most importantly the dramatic improvement of signal conditioning and data logging led to the replacement of many mechanical transducers by their electrical counterparts.

This is not to suggest that the demise of the mechanical transducer is around the corner, indeed nothing could be further from the truth. It is more that modern technological requirements need both capabilities.

The modern range of electrical transducers is wide-reaching and nearly every require-ment can be met by some device; albeit on occasions using great cunning and considerable lateral thinking.

However there may well come a time when even after the most diligent search a decision will be made to build within one's own resources a special transducer. Whatever the reason for proceeding in this manner there are basic rules to ensure that a happy conclusion is reached. This section sets out to provide such rules.

Most likely the device to be built will form an integral part of a piece of new test equipment or there will be a need to alter or modernise some existing equipment, to provide an electrical output signal. In either of these cases the size and shape is invariably dictated by the existing equipment and the transducer must be tailored to accommodate certain specific and unalterable requirements.

Unless the purpose is to produce some new technique or a unique adaptation of an old one, the sensing element is best purchased and adapted by incorporation into that part of the transducer which creates the environment suitable for the sensing device to provide a desirable response.

Before proceeding, one can identify certain transducers which should be considered as taboo for in-house development unless truly there is no other option. It is suggested that acceleration and velocity of solid bodies are unsuitable since although developing a spring/mass system is relatively easy, providing the required damping is a sophisticated task and it will be very difficult to better those transducers available commercially however that "better" quality is specified. To this may be added "Temperature" since there are already sufficient commercial devices with a wide range of size and desired accuracy to suit the most stringent requirement.

Transducers may be subject to external regulations, such as health standards or intrinsic safety regulations. Since this may involve considerable testing to satisfy some regulatory authority it is likely to inhibit all but the most determined researcher.

However, there are occasions where the needs cannot be met by commercial devices. Space availability, environmental requirements, the need for extremely high frequency response and even security – both commercial and military – may suggest proceeding with a customised design. However there are rules which are worth following and these are offered below.

Before proceeding there is one question which must be answered with care and honesty. *Is the measurement necessary?*

Often a measurement programme is contemplated and started and yet careful consideration of the evidence will either negate the need or result in a more simple solution. It is never wise to rush into making a measurement until all of the options have been considered.

As an example, consider a structural member failing in tension. Examination of the break identifies the failure as a simple overload, rather than one caused by fatigue. A knowledge of the material will identify the maximum stress and hence the resulting failure load. Measurement of the strain in service will not provide any better information. This is a simple example, taken from an actual case history, but there are many such situations where careful examination of the existing evidence will provide the solution.

Guidelines for in-house design

1. Be sure that there is no commercial device available.

Scientifically trained staff enjoy a challenge! Thus there is a tendency to make any market search rather cursory with the answer that "nothing is available" offered well in advance!

There is an extremely wide choice of transducers available to the would-be purchaser. Further there are companies who specialise in building devices to specification. This expertise should be used wherever possible. Remember that they are aware of all of the dangers which are lurking, waiting to surprise the newcomer, and the purchaser usually spends not a penny until he has a working device.

Sometimes it is possible to adapt a transducer to perform what at first may seem an unrelated task. As an example, a small load cell can be readily adapted to measure pressure, displacement, extension or flow. One example identified in the main text for flow devices uses temperature measurement as the sensor; not an obvious solution to the problem. (It is essential to remember that if a load cell is used to determine displacement the output is a measure of force however we identify it and thus is prone to all of the external influences of a force transducer as distinct from a displacement transducer.)

2. Assess the skills available.

The choice of sensing element may be determined more by the skills which are available within a company rather than the technical merit of the sensing element. An LVDT installation is mainly about mechanical design whereas the bonding of resistance strain gauges are closely linked to a wireman's skill.

It is important to accept the notion that there is invariably more than one solution to a problem and thus the act of giving a problem to one department rather than another will have considerable influence upon the outcome; in its solution, the degree of sophistication and the final cost.

3. Seek expert advice.

Suppliers and manufacturers often provide design information, from data sheets to books, on various types of transducer, some of these being listed in the Bibliography. Short courses are readily available to extend both technical and practical skills.

It may be advisable to call in a consultant for a few days to advise on possible solutions. Their experience, knowledge and independence will ensure a proper course of action. However Rule 5 below must be adhered to so that the problem is properly defined.

4. Never assume that you can produce something better than is commercially available.

There is a tendency in the scientific and engineering world to suggest that if a manufacturer can produce something to an accuracy of 0.01% it should be possible to improve this to 0.005% by customising. Rarely is this true. It is often overlooked that a commercial product is the result of years of research and development, undertaken by staff with many more years' experience. If the accuracy is truly the essential feature, talk to the manufacturer's technical sales department. They will almost certainly be able to help. Remember too that their description of errors may be such that there is the potential to restate these more realistically in relation to your requirements. As an example if one is willing to accept the output of a transducer as being nonlinear most transducer specifications can be improved considerably.

5. Prepare a written requirement.

However confident one is that everyone knows the exact requirement, seldom is this the case. In documenting the specification, be careful not to describe the likely solution but to identify the need.

A common mistake is to over-specify the quality of a measurement particularly in terms of the performance for nonlinearity, hysteresis and repeatability, and yet give little thought to, say, "survivability". A transducer is valueless if its working life may be described as "momentary". Over-specifying is often the lazy solution when identifying the desired requirement.

Be aware that the output, in itself, may not be the final result. For example if a "1%" transducer result is entered in a calculation where its output is part of a squared term, then the error will now exceed 2%, or 9% for a cubic function. If this is an unacceptable error value the whole programme may need reconsideration and the chosen measurement changed for another solution.

Included within the requirement is the need to determine the development cost. Pricing is never easy and there is a natural tendency to underestimate, particularly if there is a belief that management may abandon the project before it starts if the price is unacceptably high.

It is better to abandon any project at the beginning rather than later and engineers have no automatic right to spend money without the related financial discipline.

6. Ensure that only one person has overall responsibility for the project.

The saying about the camel being a horse designed by a committee, if rather unfair to a most remarkable animal, should none the less be heeded. However, there is nothing inherently wrong with design by committee as long as decisions are not compromises of personality. There will be enough compromises to be made in the design itself.

However it is impossible to produce transducers if there is uncertainty over who has the final decision. As this book has shown, for any particular measurement requirement there are a number of different techniques which may provide a satisfactory performance but only one choice can be made.

An important corollary is that a satisfactory solution is as much about the enthusiasm to succeed as it is on the initial choice of a possible solution. There are many transducers, some listed in this book, where only enthusiasm and dedication have ensured a satisfactory solution against the odds.

7. Identify the method of calibration.

It may seem strange that this requirement is mentioned but, during the development, calibration will become a necessity. This may include establishing the performance of parameters other than that of the transducer measurement, such as temperature sensitivity. The technique and quality of calibration will determine the best possible transducer performance.

Although it may be intended to finalise the quality of performance at a suitably-equipped external laboratory, there will be many occasions where internal checks are required. Ensure that the behaviour of this equipment is known. Its performance and that of the transducer will be indistinguishable and thus attempts to improve the performance of the transducer may be negated by the behaviour of the calibration device.

8. Set a time limit for the whole operation.

Development of any piece of equipment is never complete. There is always some way of improving the product. A new material, a better form of cable, or a new welding technique are just some of the ways in which it is possible to retard the progress in mid-activity. Without a time scale it is possible to become trapped by seeking higher degrees of perfection.

9. Prepare a programme of work.

At an early stage the likely solution should become clear. The design team should have been identified together with those undertaking other aspects of the programme. In keeping with the time scale, identify the work distribution and responsibility of each member of the team. Keep notes of all agreed actions. (This will save recriminations later.)

10. Allow the design process to evolve naturally.

Don't rush into a decision without supporting information. This is particularly important at the beginning of the design phase. The earlier decision will soon become irreversible and thus attempts to make decisions too rapidly at this point may come back and haunt the project from there on.

11. Know when to stop.

It is seldom appreciated that any project only has a certain percentage chance of success. There is no guarantee of success and it is best to admit this at the start rather than end up making excuses. There are often a number of valid reasons for abandoning the project. Accept them when the moment comes.

12. Keep it simple.

The more complex a transducer, the more difficult it is to understand its behaviour. This is bearable if it works, but when there are problems the solutions may not be so obvious. Be wary of proceeding if the solution becomes complicated.

13. Believe in Hooke's Law.

The design of the vast majority of transducers requires that there is some change in length or a deformation. Except in the most unusual circumstances this deformation is remarkably linear. Indeed it is extremely difficult to create a nonlinear response unless it is inherent within the concept, such as a vibrating wire transducer. This linearity allows the designer to take considerable liberties allowing more simple and cheaper devices to be constructed.

14. Beware of friction.

"Bearings will not transmit a moment." "Splined shaft connections slide." "Bolted structures are rigidly linked." Such statements are commonly accepted concepts in normal engineering but are not true for transducers. The errors which arise from ignoring what at first seem to be insignificant effects may cause drastic problems. Worse, friction and these other undesirable features can be and often are very repeatable in their performance and thus difficult to identify and isolate.

15. Fault finding: look carefully at all of the information.

During the development stage problems may well arise. The transducer will produce an output as will other sources, such as the calibration inputs, which identify that problem. The skill is to translate the message into a meaningful one.

One must never dismiss any part of the total information because it is not understandable or is not believable. Almost certainly the fault will be relatively simple even for the most confusing sets of results. Once the answer is found it may well be so obvious that one will wonder why it took so long to identify. In addition it is very, very seldom that what may appear to be conflicting information is due to two separate problems. To study the presented information is like being the detective in a well-written mystery story. The information is available, it just requires understanding, but of course it is not possible to skip to the back page to find the answer!

16. "Log it".

One of the most important aids to development is to record all of the relevant information. What is relevant? In simple terms, everything. Logging used to be a commonplace discipline which ensured that everything was written into hardback notebooks. This procedure is now not so prevalent. The use of loose sheets of paper is to be avoided since they are easily lost. Each entry must include the time and date. Whatever the method, record what is being done, with what, by whom and where and include full identification of the equipment.

Often the detailing of the diary of events will seem boring and pointless. Indeed it is not until there are problems that the information becomes important and needs to be studied in greater detail.

17. Don't modify the prototype.

Hopefully at the end of the project there will be a satisfactory result. This transducer may well incorporate a number of surplus or unsatisfactory features which are the result of modifications during development and it may be decided to build a new "clean" unit.

At this stage it is often tempting to incorporate other features or modifications which have not been tested earlier. Perhaps wire size or a different casing material is used. Whatever change is made, to alter a prototype is unwise. Each new feature may introduce undesirable effects and make further development necessary. Stay as far as possible with the design which has been tried and tested

18. Trust the results.

It is not uncommon that the final piece of equipment replaces some other older device. This older device may have been unreliable, have required considerable maintenance, have taken up too much operator time and may have failed to provide the desired information in the preferred format. However the moment that the replacement is installed, the old unit may be viewed by operators in a completely different way: "it WAS easy to use", "it WAS easy to repair" and so on. Do not get frustrated or annoyed at these comments. After all when the new is replaced with something even more advanced it too will receive the same treatment as things past!

19. Setting up a committee.

Many organisations require a formal committee structure when a project of reasonable size is being set up. A committee as such is not a bad thing but it must be clear as to its function. Membership must be limited to those directly connected with the project: provider or user.

The title of a committee tends to affect the way in which it operates. A group called "XYZ design team" will operate differently to "XYZ managing committee". Titles, for reasons which are not clear, have enormous influence over the activities and actions of groups of people!

20. Are you really sure that you must undertake this task?

So we return to the very first rule. It cannot be stressed too highly that in-house design may result in chaos. Are you ready for this? If not, don't start!

Many of these guidelines will seem either obvious or unimportant. It is true that it is difficult to persuade anyone that any given procedure is better than another. However all of the above has, for the author at least, withstood the test of time, and has resulted in many more successes than failures.

However you choose to proceed, remember that there must be organisation and proper control. Without it success will be a matter of chance with the odds for completion being very poor.

All success to your endeavours!

Bibliography

Transducer technology, as a scientific subject, is relatively poorly served with written material. Most of the information comes in the form of papers, which very often describe some new development rather than the state of the art. Hence in preparing this bibliography it has been necessary to identify a range of publications of a mixed nature including many from the manufacturers' literature, which centre upon particular products but often contain useful general information.

All have been read by the Author; indeed some have been constant companions. Many of them contain further references to aid the seeker of knowledge. The titles can be misleading and thus appropriate comments are made to identify the nature of the subject and the benefit arising from reading the information. This is not an attempt to provide a balanced reference source. The literature listed is a selection from which the Author has benefited in his career or which he has found useful in the preparation of this book.

The listings are grouped in alphabetical order as follows:

Acceleration

Capacitive devices

Force

General

Gyroscopes

Inductive devices

Piezo-electric devices

Pressure measurement

Stress and strain

Temperature

Acceleration

Accelerometer Instruction & Selection Manual published by Entran International.

This booklet describes in detail the operation and performance of the semiconductor strain gauge accelerometers. Whilst limited to the manufacturer's own products, for those who seek an introduction to the mysteries of acceleration this booklet provides a good introduction with the bonus of information on the use of semiconductor strain gauges.

Capacitive devices

High Accuracy Capacitive Transducers by P. Wolfendale and B. Cropper and available from Automatic Systems Laboratories Ltd.

This is a brief (five pages) and succinct tract, including a brief resumé of the manner in which capacitance change can be engineered. Potential accuracies and likely sensitivity, resolution and stability are discussed. Further it identifies particular hostile environments which may be tolerated by the capacitance transducer.

Force

Load Cell Instruction & Selection Manual published by Entran International.

This booklet describes in detail the operation and performance of the semiconductor strain gauge load cells. Although limited to the manufacturer's own products its contents do give much useful information on both design of load cells and the proper use of semiconductor strain gauges.

General

Machinery's Handbook published by Industrial Press Inc.

My copy, the 22nd Edition (1985), indicates that new editions have been produced on a regular basis since 1914. Although this book is often to be seen in mechanical design offices and workshops it is sometimes forgotten that there is a significant sized section on Strength of Materials, and it contains an enormous amount of information useful to the transducer engineer.

Reports in Applied Measurements published by Hottinger Baldwin Messtechnik GmbH.

This free Journal is published twice yearly and is available from HBM. Each issue contains a number of papers covering some measurement task or programme of work. Although these papers highlight the use of HBM equipment the various authors are identified and are independent of HBM. A good read for any measurement engineer.

Gyroscopes

Gyro Guide for Systems Engineers by Humphrey Inc. available from them or IFO International (UK) Ltd.

Gyroscopes and rate sensors are often a closed book to many engineers. This open book, written by a manufacturer of such devices, allows the reader to start his education and is a good introduction to the various terms used in this specialisation.

Inductive devices

Handbook of Measurement & Control by Edward E. Herceg and published by Schaevitz Engineering (Lucas Schaevitz).

Specialising on the theory and application of the LVDT, it includes many measurement applications such as force, pressure and displacement. Further it provides an authoritative work on measurement including chapters on "An Introduction to Measurement" and "Electromechanical Systems".

Piezo-electric devices

Piezo-electric Measuring Instruments and their Applications by R. Kail and W. Mahr, available from Kistler Instruments Ltd.

This paper is translated from the Article *Piezoelektrische Messgeräte und ihre Anwendungen* edited in *Messen und Prüfen* Volume 20 (1984) 7...12. It is essential reading for those who wish to understand the piezo-electric device. In addition to explaining the properties of quartz crystal, aid is given in the practical design of transducers. There are sections on signal conditioning, frequency response, calibration and typical applications with a wealth of technical information.

The bibliography contains 27 references, the majority in German. Kistler offer a reprint service.

Piezo Guide: Piezo Positioning Technology, Parts 1 and 2 by Physik Instrumente and available from LPL Lambda Photometrics Ltd.

Although these are the manufacturer's product catalogues, they include considerable detail about the operation of these devices. It includes the various formulae which identify, for example, both static and dynamic performance, heating effects, temperature response and so on.

Pressure measurement

Instability Problems with Absolute High Pressure Transducers used for Seabed Pressure Measurements Volume 1 of the Test and Transducer Conference, October 1986.

The title of this paper suggests a specialised application. The text however gives insight into the performance and long-term stability of three different types of pressure transducer – piezoresistive, integrated silicon sensor and a quartz resonator, which have wider applications than seabed use alone. The paper discusses temperature correction, creep compensation and stability both in use and in the laboratory, and certainly is worthwhile reading for anyone engaged upon precision long-term pressure measurement.

Pressure Transducer Instruction & Selection Manual published by Entran International.

This booklet describes in detail the operation and performance of diaphragm style pressure transducers using bonded semiconductor strain gauges. Whilst limited to the manufacturer's own products its contents are a good primer on the subject of pressure with the bonus of information on the use of semiconductor strain gauges.

Stress and strain

An Introduction to Measurements using Strain Gauges by Karl Hoffmann published by Hottinger Baldwin Messtechnik GmbH.

Translated from an original version in German, this book is a recent text on an important subject. It is a most comprehensive book and although one is surprised by the occasional omission, such as no mention of vibrating wire devices, one can overlook these limitations since within its 291 pages there is a remarkable amount of information; from "Adhesives" to "Zero Referenced Measurements" as its index identifies. Valuable reading for both the beginner and expert.

Applications & Installation Manual prepared by the Eaton Corporation and available from them.

This manual is a dissertation on the Eaton weldable and embeddable strain gauges. Essential reading when the use of this type of strain gauge is being contemplated.

Basic Stress Analysis by M J Ironmonger published by Butterworth – Heinemann Ltd.

This book exploits simple BASIC computer programmes to aid the calculations required of many stress analysis applications. Since many of the examples, complete with the appropriate computer programme listings, are related to beams under various forms of loading they have direct relevance to many transducer designs. For any would-be transducer designer this book offers a starting point for the use of a computer to simplify the calculation process.

BSSM Strain Measurement Reference Book prepared by J. Pople and published by The British Society of Strain Measurement.

This book identifies and describes all of the normal strain measurement techniques although the majority of the information is concentrated on the bonded resistance strain gauge. It is essential reading for those who wish to have a better understanding of the installation and use of the strain gauge whatever the final purpose.

Handbook of Experimental Stress Analysis edited by M. Hetenyi and published by John Wiley (USA) and Chapman & Hall Ltd (UK).

This book is now out of print but still offers interesting reading and inspiration. Although dedicated to stress analysis it does provide information on testing machines, mechanical gauges and extensometers, optical methods, inductive and capacitive gauges and motion measurement; much of which is as relevant today as at the time of its first printing (circa 1950).

This book is now replaced by:

Handbook on Experimental Mechanics edited by Albert S. Kobayashi and published by SEM Publications Dept, 7 School St, Bethel, CT06801, USA.

Since the publication of Hetenyi's book techniques have developed and evolved. This new book reflects those changes and is a worthy successor to its counterpart. New sections include holography, interferometry, moiré and modal analysis as well as updated information of the more traditional techniques.

Handbook of Formulas for Stress & Strain by William Griffel and published by Frederick Ungar Publishing Co.

A large book of over 400 pages, it is crammed with formulae, tables and graphs. Its title is an accurate description of the book and with its tables and nomograms it is easy to use. A most useful book for transducer designers and users alike. It provides solutions to many unexpected design criteria and could well be the source information when evolving appropriate computer programs.

Measurements Group Tech Notes published by Measurements Group Inc.

This is a collection of technical notes with each note concentrating upon some particular aspect of the use of bonded metallic foil strain gauges and temperature sensors. Although the majority of the Notes concentrate upon gauge performance, some delve into transducer applications such as Tech Note TN-510, *Design Considerations for Diaphragm Pressure Transducers.*

Strain is the Journal of the British Society for Strain Measurement, published by BSSM.

This Journal comes free to members of BSSM and is essential reading for all interested in stress, strain and experimental mechanics. Further its contents will be of great interest to all measurement engineers.

SR-4 Strain Gauge Handbook prepared by BLH Electronics and available from them or Nobel Systems Ltd.

A reasonably detailed source of information on the use of bonded metallic strain gauges both for strain measurement and for other transducer applications.

Strain Gage Based Transducers by the Technical Staff of Measurement Group, Inc.

This book concentrates upon the design of strain gauge load cells. However it provides invaluable information on such topics as gauge selection, compensation networks, bonding and transducer materials. Thus it is useful for any type of transducer where resistance strain gauges are likely to be used as the sensing elements.

Telemetry Applied to Strain Gauging by K. J. Phillips. Published in Strain, February 1985 and available from Adretta Ltd.

This paper describes the advantages of telemetry systems over other strain gauge signal capture and transmission systems. It includes a comparison between frequency modulation and pulse duration modulation techniques.

Kulite Semiconductor Strain Gage Manual, published by Kulite Semiconductor Products Inc.

For anyone who wishes to further their knowledge of semiconductor strain and temperature gauges, this booklet is essential reading. It covers in detail all of the important aspects of these devices and is supported by appropriate mathematics where necessary.

Temperature

Infrared Noncontact Temperature Measurement published by Calex Instrumentation.

As the front cover states this is their Temperature Handbook and Catalogue and the handbook part discusses remote temperature sensing in great detail. It is full of useful information on thermocouples in general and includes tables of the thermo-electric outputs for all of the commonly used thermocouple types.

Surface Temperature Measurement by Randal A Gauthier. Magazine article based upon a presentation by the author at the 1980 ISA International Conference, available from Rhopoint Ltd.

A useful description of the choice, installation and use of resistance temperature detectors (RTD) including some case histories.

Temperature and Heat Flow Measurement using Foil Devices by P. G. Brown, Test & Transducer Conference, October, 1986, Volume 1 and available from Rhopoint Ltd.

This describes the practical problems and solutions of measuring the temperature of a surface using both platinum resistance thermometers and thermocouples. In addition it discusses the use of heat flow sensors and provides details of their calibration.

Appendix – List of Suppliers

The following list of companies are those from whom most of the technical information included within this book has been obtained. They are identified by their codes within the text under the heading of "Suppliers".

Although the information was correct at the time of preparation, experience shows that company names, addresses, type of products and telephone numbers do change more frequently than one would expect.

Where companies have an international organisation and the information was available, these too have been listed.

Where no country of origin is given, the address is in the United Kingdom.

ABB1   ABB Kent-Taylor Ltd

Oldends Lane, Stonehouse, Gloucestershire GL10 3TA

Tel: 045382 6661; Fax: 045382 6358

ABB2   ABB Process & Automation Ltd

Unit 19/20 Gemini Business Park, Europa Boulevard, Warrington, Cheshire WA5 5TU

Tel: 0925 232900

ABB3   ABB Automation

ABS-721, 67 Vasteras, Sweden

Tel: 021-10 90 00; Fax: 021-11 93 48

ACT   Active Load Ltd

Millstone Wood Lane, Woodcote, Reading, Berkshire RG8 0PU

Tel: 0491 680123; Fax: 0491 680123

ADR   Adretta Limited

North Street, Wellington, Somerset TA21 8LZ

Tel: 0823 663535; Fax: 0823 663535

AIR   Airflow Developments Ltd

Lancaster Road, Cressex Industrial Estate, High Wycombe, Buckingham-shire HP12 3QP

Tel: 0494 525252; Fax: 0494 461073

AJB   AJB Associates Ltd

54 High Street, Wells, Somerset BA5 2SN

Tel: 0749 74932; Fax: 0749 670265

AKG1   AKG Accoustics Ltd

Vienna Court, Catteshall Wharf, Cat-teshall, Godalming GU7 1JG

Tel: 0483 425702; Fax: 0483 428967

AKG2   AKG Akustiche U Kino-Gerate GmbH

Brundhildengasse 1, Wein A-1150, Austria

Tel: 222 956517

ANL   Anley Controls Ltd

16 Anley Road, London W14 0BY

Tel: 071 6034007; Fax: 071 3712172

APE Apek Sales Ltd

6 Haviland Road, Ferndown Industrial Estate, Wimborne, Dorset BH21 7RF

Tel: 0202 877044

APP Apparecchi Di Controllo Ranco

Via Del Seprio 42, 22074 Lomazzo (Como), Italy

Tel: (02) 9637 0358

AUT Automatic Systems Laboratories Ltd

28 Blundells Road, Bradville, Milton Keynes, Bucks MK13 7HF

Tel: 0908 320666; Fax: 0908 322564

BIR BIRAL (Bristol Industrial & Research Associates Ltd)

PO Box 2, Portishead, Bristol BS20 9JB

Tel: 0275 847787; Fax: 0275 847303

BRI Bristol Babcock Ltd

Parsonage House, Parsonage Square, Dorking, Surrey RH4 1UP

Tel: 0306 740033; Fax: 0306 740603

BRO Brookes & Gatehouse Ltd

International Marine House, Abbey Park, Romsey, Hants SO51 9AQ

Tel: 0794 518448; Fax: 0794 518077

BRU1 Bruel & Kjaer (UK) Ltd

Harrow Weald Lodge, 92 Uxbridge Road, Harrow HA3 6BZ

Tel: 081-954 2366

BRU2 Bruel & Kjaer A/S,

Naerum DK-2850, Denmark

Tel: 2 800500

RU3 Bruel & Kjaer Instruments Inc

185 Forest Street, Malborough MA 01752, USA

Tel: (617) 481-7000

BUD Budenberg Gauge Co. Ltd

P.O. Box 5, Altrincham, Cheshire WA14 4ER

Tel: 061 9285441; Fax: 061 9287075

CAL Calex Instrumentation Ltd

PO Box 2, Leighton Buzzard, Bedfordshire LU7 7AG

Tel: 0525 373178; Fax: 0525 851319

CAM1 Camille Bauer Controls Ltd

Priest House, Priest Street, Cradley Heath, Warley, West Midlands B64 6JN

Tel: 0384 638822; Fax: 0384 639168

CAM2 Camille Bauer Ltd

CH-5610 Wohlen, Switzerland

Tel: (0)57 212111; Fax: (0)57 227432

CON Concepts (Laleham) Ltd

204A Thames Side, Laleham, Middlsex TW18 2JN

Tel: 0784 455695; Fax: 0932 248600

DES Desoutter Limited

319 Edgeware Road, Colindale, London NW9 6ND

Tel: 081-205 7050; Fax: 081-205 5167

DEU Deutsche Ranco GmbH

Am neuen Rheinhafen 6720, Speyer, Germany

Tel: 06232-120

DRU Druck Limited

Fir Tree Lane, Groby, Leicester LE6 0FH

Tel: 0533 314314; Fax: 0533 875022

END1 Endevco (UK)

Melbourn, Royston, Herefordshire SG8 6AQ

Tel: 0763 261311; Fax: 0763 261120

END3 Endevco

30700 Rancho Viejo Road, San Juan, Capistrano, California 92675 USA

Tel: (714) 493-8181

ENT Entran Ltd

Entran House, 19 Garston Park Parade, Garston, Watford, Herts WD2 6LQ

Tel: 0923 893999; Fax: 0923 893434

EUR1 Eurisem Technics

40 High Street, Earl Shilton, Leicestershire LE9 7DG

Tel: 0455 848424; Fax: 0455 840552

EUR2 European Instruments

The Trading Estate, Old Road, Headington, Oxford OX3 8TA

Tel: 0865 750375; Fax: 0865 69985

EUR3 European Monitoring Services

Unit 208, Solent Business Centre, Milbrook Road West, Southampton

Tel: 0703 788555; Fax: 0703 785111

EXP Explorocean Technology Ltd

Unit 2, Pool Road, West Molesey, Surrey KT8 0HE

Tel: 081 9413656

GAE Gaeltec Ltd

Dunvegan, Isle of Skye, Scotland

Tel: 047 022385; Fax: 047 022369

GEC GEC Avery Ltd

Smethwick, Warley, West Midlands B66 2LP

Tel: 021558 1112; Fax: 021565 6062

GIL Gill Instruments Ltd

Solent House, Cannon Street, Lymington, Hampshire SO41 9BR

Tel: 0590 671754; Fax: 0590 676409

GRA Graham & White Instruments

135 Hatfield Road, St Albans, Hertfordshire AL1 4LZ

Tel: 0727 59373; Fax: 0727 44272

HBM1 HBM (UK) Ltd

Station Approach, Bicester, Oxfordshire OX6 7BT

Tel: 0869 321321; Fax: 0869 321111

HBM2 HBM GmbH

Im Tifen See 45, 6100 Darmstadt 1, Germany

Tel: (0)6151/32-0; Fax: (0)6151/893686

HUB Hubner Elektromaschinen

AGD-1000 Berlin 61, Postbox 610271, Planufer, Germany

Tel: (030)690 03-0-104

HUM Humphrey Inc.

9212 Balboa Avenue, San Diego, California, USA

Tel: 619 565-6631

HYD Hydraulics Research Ltd

Wallingford, Oxfordshire OX10 8BA

Tel: 0491 35381; Fax: 0491 32233

IFO   IFO International (UK) Ltd

3 Beechwood Drive, Keston, Kent BR2 6HN

Tel: 0689 856768; Fax: 0689 855933

INT1   Interface Engineering

Marina Building, Harleyford, Marlow, Buckinghamshire SL7 2DX

Tel: 06284 74248

IRA   Irad Gage

Etna Road, Lebanon, NH 03766, USA

Tel: 603-448-4445

KDG   KDG Mobrey Ltd

190-196 Bath Road, Slough, Berkshire SL1 4DN

Tel: 0753 534646; Fax: 0753 823589

KIS1   Kistler Instruments Limited

Whiteoaks, The Grove, Hartley Wintney, Hampshire RG27 8RN

Tel: 025126 3555; Fax: 025126 4439

KIS2   Kistler Instrument Corp.

75 John Glenn Drive, Amherst, NY 14120, USA

Tel: (716) 691-5100; Fax: (716) 691-5226

KIS3   Kistler Instrumente GmbH

Friedrich-List-Strasse 29D-7302, Ostfildem 2, Germany

Tel: (0711) 3407-0; Fax: (0711) 3407-159

KIS4   Kistler Instruments

AGCH-8408 Winterthur, Switzerland

Tel: (052) 831111; Fax: (052) 257200

KRO   Krohne Measurement & Controls

Osyth Close, Brackmills, Northants NN4 0ES

Tel: 0604 766144; Fax: 0604 769677

KUL1   Kulite Sensors Ltd

Kulite House, Stroudley Road, Kingsland Business Park, Basingstoke, Hants RG24 0UG

Tel: 0256 461646; Fax: 0256 479510

KUL2   Kulite Semiconductor Products Inc.

1039 Hoyt Avenue, Ridgefield, New Jersey, USA

Tel: 201 945 3000; Fax: 201 943 3294

LAB   Labfacility Ltd

99 Waldgrave Road, Teddington, Greater London TW11 8LR

Tel: 081-943 5331; Fax: 081-943 4351

LIT   Litre Meter

50/53 Rabans Close, Rabans Lane Industrial Estate, Aylesbury, Bucks HP19 3RS

Tel: 0296 20341; Fax: 0296 436446

LPL   LPL Lambda Photometrics Ltd

Lambda House, Batford Mill, Harpenden, Herfordshire AL5 5BZ

Tel: 05827 64334; Fax: 05827 2084

LUC1   Lucas Schaevitz Ltd

543 Ipswich Road, Slough, Berkshire SL1 4EG

Tel: 0753 537622; Fax: 0753 823563

LUC2   Lucas Sensing Systems Inc.

21640 N. 14th Avenue, Phoenix AZ, 85027-2839, USA

Tel: (800) 545-3243; Fax: (602) 582-3520

MAR   Marsden Weighing Machine Group Ltd

388 Harrow Road, London W9 2HU

Tel: 071 2891066

MAY Maywood Instruments

Rankine Road, Daneshill Industrial Estate, Basingstoke, Hampshire RG24 0PP

Tel: 0256 57572; Fax: 0256 840937

MEA1 Measurements Group (UK) Ltd

Stroudley Road, Basingstoke, Hampshire RG24 0FW

Tel: 0256 462131; Fax: 0256 471441

MEA2 Measurements Group Inc

PO Box 27777, Raleigh, NC 27611, USA

tel: (919) 365-3800

MEC Mecmesin Ltd

Unit 3, Lawson Hunt Industrial Park, Broadbridge Heath, West Sussex RH12 3JR

Tel: 0403 56276; Fax: 0403 210106

MHH MHH Engineering Co. Ltd

Torqueleader, Bramley, Guildford, Surrey GU5 0AJ

Tel: 0483 892772

MIC1 Micro-Epsilo Messtechnik

6 Hall Park Hill, Berkhamstead, Herts HP4 2NH

Tel: 0442 877991; Fax: 0442 877992

MIC2 Micro-Epsilon Messtechnik GmBH & Co.

KG Konigbacher Strasse, 15D-8539 Ortenburg-Dorfbach, Germany

Tel: 0 8542/7978

MIC3 Micron Instruments Inc.

1519 Pontius Avenue, Los Angeles, California, 90025 USA

Tel: 213/478-0964

NAS NASA (1992) Ltd

Boulton Road, Stevenage, Hertfordshire SG1 4QG

Tel: 0438 354033; Fax: 0438 741498

NAT National Physical Laboratory

D.M.O.M., Teddington, Middlesex TW11 0LW

Tel: 081 977 3222; Fax: 081 943 2155

NEG Negretti Automation Ltd

Stocklake, Aylesbury, Buckinghamshire HP20 1DR

Tel: 0296 395931; Fax: 0296 431435

NOB Nobel Systems Ltd

Murdock Road, Bedford, Bedfordshire MK41 7PQ

Tel: 0234 349241 0234 325387

NOR Norbar Torque Tools Ltd

Beaumont Road, Banbury, Oxfordshire OX16 7XJ

Tel: 0295 270333; Fax: 0295 269864

OCE Oceano Instruments (UK) Ltd

9 New Broompark, Granton Park Avenue Industrial Estate, Edinburgh EH5 1RS

Tel: 031 552 0303; Fax: 031 552 6619

PEN Penny & Giles Ltd

8 Airfield Way, Christchurch, Dorset BH23 3TT

Tel: 0202 499171

PHY Physik Instrumente GmBH & Co

Siemensstr, 13-15D-7517 Waldbronn, Germany

Tel: 07243 604-0; Fax: 07243 69944

PIA PIAB Ltd

PO Box 8, Brentford, Middlesex TW8 9AJ

Tel: 081 568165

PLA Platon Instrumentation Ltd

Platon Park, Viables, Basingstoke, Hampshire RG22 4PS

Tel: 0256 460122; Fax: 0256 63345

POW Powertronic International

Blenheim Barn, Tidmarsh Lane, Pangbourne, Berkshire RG8 8HH

Tel: 0734 845351; Fax: 0734 843979

PRE Precision Varionics Ltd

Units 8 & 9, Cwmdraw Industrial Estate, Ebbw Vale, Gwent NP3 5AQ

Tel: 0495 350522; Fax: 0495 350533

RAL Ralcom Automation Ltd

The Old Mill, 61 Reading Road, Pangbourne, Reading, Berks RG8 7HZ

Tel: 0734 845285 Fax: 0734 844114

RAN1 Ranco Controls Ltd

Southway Drive, Southway, Plymouth PL6 6QT

Tel: 0752 777166

RAN2 Ranco France S.A.

Z.I. 38 Rue Senouque, 78530 BUC., France

Tel: (3) 956 50 63

RDP RDP Group

Grove Street, Heath Town, Wolverhampton WV10 0PY

Tel: 0902 57512; Fax: 0902 52000

REN Renishaw Metrology Ltd

New Mills, Wotton-under-Edge, Gloucestershire GL12 8JR

Tel: 0453 524524; Fax: 0453 524102

RHO Rhopoint Ltd

Holland Road, Hurst Green, Oxted, Surrey RH8 9BB

Tel: 0883 722222; Fax: 0883 717245

RMY R.M. Young Co

2801 Aero-Park Drive, Traverse City, Michigan 49684, USA

ROC Rock Instruments Ltd

Bell Lane, Uckfield, East Sussex TN22 1QL

Tel: 0825 765044; Fax: 0825 761740

ROS Rosemount Engineering Co. Ltd

Heath Place, Bognor Regis, Sussex PO22 9SH

Tel: 0243 863121; Fax: 0243 867554

RSC RS Components Ltd

PO Box 99, Corby, Northants NN17 9RS

Tel: 0536 201234; Fax: 0536 201501

SAN Sandhurst Scientific Instrument Co. Ltd

25 Perry Hill, Sandhurst, Camberley, Surrey GU17 8HS

Tel: 0252 875131

SAR1 Sartorius Instruments Ltd

Long Mead Business Centre, Blenheim Road, Epsom, Surrey KT19 9QN

Tel: 0372 745811; Fax: 0372 720799

SAR2 Sartorius AG

P.O. Box 32433400, Goettingen, Germany

Tel: 0551 308-1; Fax: 0551 308 289S

SCA1 Scaime Techno Talent

B.P. 501-Z.I., De Juvigny, 74105 Annemasse, Cedex, France

Tel: (33)50877864; Fax: (33)50877946

SCA2 Scaime - PTC

P.O. Box 72, Wycoff, NJ 07481, USA

Tel: 201 652 84 94; Fax: 201 652 31 86

SEN1 Sensit Ltd

Cambell Court, Bramley, Basingstoke, Hampshire RG26 5EG

Tel: 0256 882992; Fax: 0256 882986

SEN2 Sensor Technology

Balscote Mill, Banbury, Oxfordshire OX15 6JB

Tel: 0295 730746

SEN3 Sensortechnics

30 Regent Place, Rugby, Warwickshire CV21 2PN

Tel: 0788 560426 0788 561228

SEN4 Sensortechnics GmbH

HD-8039-Puchheim, Germany

Tel: (089) 800 830

SHI Ship & Marine Data Systems

Units 3 & 4, 20 Union Road, Kingsbridge, Devon TQ7 1EF

Tel: 0548 852552; Fax: 0548 852112

SMI Smith Industries Ltd

Bishops Cleeve, Cheltenham, Gloucestershire GL52 4SF

Tel: 024267 3333

SOI Soil Instruments Ltd

Bell Lane, Uckfield, East Sussex TN22 1QL

Tel: 0825 765044; Fax: 0825 761740

STR Strainstall Engineering Services Ltd

Denmark Road, Cowes, Isle of Wight PO31 7TB

Tel: 0983 295111; Fax: 0983 291335

TEC Techni Measure

Alexandra Buildings, 59 Alcester Road, Studley, Warrington B80 7NJUK

Tel: 052785 4103

TED Tedea-Huntleigh Europe

Portmanmoor Industrial Estate, Cardiff CF2 2HB

Tel: 0222 460231; Fax: 0222 462173

TEL Telesonic Marine Ltd

60-64 Brunswick Centre, Marchmont Street, London WC1N 8AE

Tel: 071 837 4106; Fax: 071 837 4100

TES Testoterm Ltd

Old Flour Mill, Queen Street, Emsworth, Hampshire PO1 0BT

Tel: 0243 37222; Fax: 0243 378013

THA Thames Side Scientific Co. Ltd

Southview Park, Caversham, Berkshire RG4 0AF

Tel: 0734 474379; Fax: 0734 484327

THE The Fredericks Company,

P.O. Box 67, Philmont Avenue & Anne Street, Huntingdon Valley, PA 19006, USA

Tel: (215) 947-2500; Fax: (215) 947-7464

TIL Tilt Measurement Ltd

Horizon House, London Road, Baldock, Herts SG7 6NG

Tel: 0462 894566

TIN Tinsley Strain Measurement Ltd

275 King Henry's Drive, New Addington, Croydon, Surrey CRO 0HE

Tel: 0689 800799; Fax: 0689 800405

TSI1 TSI Corporate Headquarters

500 Cardigan Road, P.O. Box 64394, St Paul, MN 55164, USA

Tel: (612)483-0900

TSI2 TSI France Incorporated

Centre d'affaires INTEGRAL, 68 rue de Paris, 93804-Epinay, Sur-Seine-Cedex, France

Tel: (1) 4823.21.31

TSI3 TSI GmbH

Schmiedstrasse, 35100 Aachen, Germany

Tel: 0241-48121

UCC1 UCC International Ltd

PO Box 20, Thetford, Norfolk IP24 3QZ

Tel: 08427 64151; Fax: 08427 65504

UCC2 UCC France

7 rue Jules Berthonneau, 41000 Blois, France

Tel: 054 20 06 88; Fax: 054 20 09 88

UCC3 UCC GmbH

Postfach 300193, Oppeiner Strasse 14050, Monchengladbach 3, Germany

Tel: 02166 60 30 31; Fax: 02166 20 09 88

VAI1 Vaisala (UK) Ltd

Cambridge Science Park, Milton Road, Cambridge CB4 4BH

Tel: 0223 862112; Fax: 0223 860988

VAI2 Vaisala Oy,

PL 26SF-00421, Helsinki, Finland

Tel: 358 0 894 91; Fax: 358 0 894 9227

VAL Valeport Marine Scientific Ltd

Unit 7, Townstal Industrial Esta, Dartmouth, Devon TQ6 9LX

Tel: 0803 834031; Fax: 0803 834320

VEC Vector Instruments

115 Marsh Road, Rhyl, Clwyd LL 2AB

Tel: 0745 350700; Fax: 0745 344206

VER Verospeed

Stansted Road, Boyatt Road, Eastleigh, Hampshire SO5 4ZY

Tel: 0703 644555; Fax: 0703 477144

WES Western Sensors Limited

The Byre, Wrexham Technology Park, Wrexham, Clwyd LL13 7YP

Tel: 0978 291731; Fax: 0978 291733